International Conference on Traffic and Transportation Engineering and Management (TTEM2019)

AA007305

IOP Conference Series: Materials Science and Engineering
Volume 792

Guangzhou, China
20 - 22 December 2019

ISBN: 978-1-7138-2670-5
ISSN: 1757-8981

Printed from e-media with permission by:

Curran Associates, Inc.
57 Morehouse Lane
Red Hook, NY 12571

CURRAN ASSOCIATES INC. proceedings.com

Some format issues inherent in the e-media version may also appear in this print version.

Institute of Physics
Dirac House, Temple Back
Bristol BS1 6BE UK

Phone: 44 1 17 929 7481
Fax: 44 1 17 920 0979

techtracking@iop.org

Additional copies of this publication are available from:

Curran Associates, Inc.
57 Morehouse Lane
Red Hook, NY 12571 USA
Phone: 845-758-0400
Fax: 845-758-2633
Email: curran@proceedings.com
Web: www.proceedings.com

TABLE OF CONTENTS

IOP Conf. Series: Materials Science and Engineering **792** (2020) 011001 doi:10.1088/1757-899X/792/1/011001

Preface

The IAECST-International Conference on Traffic and Transportation Engineering and Management (TTEM 2019), organized by Tongji University, enjoyed a big success in Guangzhou, China on December 20-22, 2019. We aim to provide a sharing platform for experts, scholars and practitioners in the field of Traffic and Transportation Engineering and Management from east and west to share their latest research achievements, explore the frontier technologies and enhance the development of Traffic and Transportation Engineering.

During the seminar, Mr. Zuguo Cheng, senior engineering from Institute of Rail Transit, Tongji University, gave his research speech on vehicles and rail transportation system. Then we also had Dr. Devinder Yadav from The University of Nottingham shared his research on *Regulatory Framework for Development of the Aviation Industry*. Finally, associate professor Bing Zhang from the school of Transportation and Logistics, East China Jiaotong Univeristy also share his view on the development of transportation engineering in China. These insightful keynote speeches shed some spotlight on the conference, which gained huge applause on site.

The proceedings of TTEM 2019 were carefully selected and reviewed under rigid process, which cover a wide variety of topics, such as traffic information engineering, traffic and transportation planning and management, road engineering and planning, traffic and transportation economy. We are pleased to present the proceedings of TTEM 2019, which are ready for international publication.

Our heartfelt thanks go to the keynote speakers and all participants attending the conference, who have made TTEM 2019 a great success. We also want to thank our sponsors for their generous supports. We hope the proceedings of TTEM 2019 can serve as a helpful reference to the readers around the world. We aim to hold this conference every year, and we hope it can be held better next time. Wish we can see you next year. Thanks a lot.

The Committee of TTEM 2019

IOP Conf. Series: Materials Science and Engineering **792** (2020) 011001 doi:10.1088/1757-899X/792/1/011001

Committee member

Speaker at the seminar

Zuguo Cheng, Institute of Rail Transit/Tongji University
Qizhou Hu, School of Automation/Nanjing University of Science and Technology (NUST)
Devinder Yadav, The University of Nottingham
Bing Zhang, School of Transportation and Logistics/East China Jiaotong University

Academic Committee

Zhiyu Chen, Research Institute of Highway Ministry of Transport
BaoHua Guo, Henan Polytechnic University
Wenbing Guo, Henan Polytechnic University
Hong Jia, Research Institute of Highway Ministry of Transport
Aizeng Li, Henan University of Urban Construction
Guangrui Tang, National Network Energy Research Institute Co., Ltd
Gang Wu, China National Institute of Standardization
Bo Wang, Traffic Management Research Institute of the Ministry of Public Security
Liang Wei, Shandong Jiaotong University
Huanhuan Yin, Research Institute of Highway Ministry of Transport
Qiuping Yang, Research Institute of Highway Ministry of Transport
Yi Zhang, Henan University of Science and Technology
Bo Zhang, Wuhan Donghu University
Honghai Zhang, Nanjing University of Aeronautics and Astronautics

Peer review statement

All papers published in this volume of *IOP Conference Series: Materials Science and Engineering* have been peer reviewed through processes administered by the proceedings Editors. Reviews were conducted by expert referees to the professional and scientific standards expected of a proceedings journal published by IOP Publishing.

IOP Conf. Series: Materials Science and Engineering **792** (2020) 012001 doi:10.1088/1757-899X/792/1/012001

An Assessment of Modes of Passenger Transportation in Beijing: Problems and Solutions

Haibin ZHAO[1*]

[1] China Urban Sustainable Transport Research Centre (CUSTReC),China Academy of Transportation Sciences, Beijing 100029, China

[*] E-mail: 358608288@qq.com

Abstract. The purpose of this study was to analyse the historical trends of various modes of passenger transportation in Beijing with respect to planning, policies, performance, and public behaviour based on statistical data. Using Beijing's passenger transportation data since about 2003, the discussion addresses changes in (1) Beijing's passenger transportation supply and demand, (2) population density, and (3) the apparent causes and effects of passenger transportation policy at various stages of policy implementation. The analysis revealed that budgetary shortfalls, air pollution, population density, and lack of governmental integration are persistent problems. Specific proposals for ameliorating those problems are presented.

1. Introduction

There is a trend in the evolution of passenger transportation in most cities in developed and large cities in developing countries. The early stage of urban expansion is primarily characterized by non-motor vehicle and public transit that gradually becomes motorized and dominated by private road vehicles, which, ultimately, in response to traffic congestion and environmental pollution, reverts to public transit and restrictions on the use of private road vehicles. Between 2003 and 2013, Beijing experienced this evolutionary process. Its population of permanent residents increased from 14.56 million to 21.15 million; daily trips within the city surged from 18.32 million to 30.33 million; private passenger road vehicle ownership grew from 2.12 million to 5.44 million; and policies regarding private passenger road vehicle ownership and use cared were taken one another [1]. A system of sharing bicycles also emerged.

Between 2003 and 2013, Beijing invested hundreds of billions of Yuan to develop its transportation infrastructure and mass transit vehicles (road, rail, and bus) to deal with traffic congestion and environmental pollution and a modal shift from private passenger road vehicles to mass transit is evident. Moreover, the use of private passenger road vehicles began to stagnate. This evolution of urban passenger transportation in Beijing is typical of cities and is being experienced in China's second-tier cities. During its evolution of passenger transportation, Beijing implemented various plans and policies that actively or passively influenced public use of the various modes of passenger transportation in the city. Assessing the effects of the city's efforts by concluding intrinsic characteristics of this process using panel data covering a long historical period is quite enlightening for latecomers.

This study (1) examines the long-term trends in the evolution of passenger transportation in Beijing, (2) evaluates the effects of transportation planning and policy on passenger transportation problems, and (3) discusses suggestions for future development. The problems and the corresponding suggestions that are addressed can help China's and other developing countries' second-tier and third-tier cities to cope with issues related to passenger transportation. These are problems that they may face during the evolutionary process and the suggestions offer far-sighted planning guidance, which may enhance the advantages of later development.

2. Background and the evolution of passenger transportation in Beijing

Beijing has experienced rapid growth over the past decade and its permanent population is increasing. In all cities, the evolution of passenger transportation is closely related to the development of social and economic dimensions of urban growth. Therefore, the analysis of the development of urban passenger transportation and the relevant policies should account for the context of the socioeconomic environment. The changes to the geography, population, and economy of Beijing must be considered as well as the relationships of these contextual factors to the evolution of passenger transportation.

2.1. Socioeconomic background

Beijing's gross domestic product (GDP) during the past decade increased dramatically each year. In 2013, the regional GDP was CNY 1950.10 billion, a 289.47% increase from CNY 500.70 billion in 2003. The per capita GDP also rapidly increased; in 2013, it was CNY 93.21 thousand, a 168.03% increase from CNY 34.78 thousand in 2003. Local fiscal revenues and expenditures also experienced rapid increases. In 2013, Beijing's public financial budget income was CNY 366.11 billion, with an estimated annual rate of increase of 10.40% [1] [2]. Its completed public finance budget spending was CNY 374.08 billion and fiscal subsidies for public transportation operations were CNY 20.01 billion. Beijing's rapid economic development has been accompanied by increased urban passenger traffic and it has simultaneously provided the financial basis of the transportation infrastructure.

Between 2003 and 2013, Beijing's permanent resident population increased 45.26% from 14.56 million to 21.15 million people. The 2013 population density was 1289 people/km^2, which is more than twice the population density in 2003 (an increase of 682 people/km^2). About 59.27% of the increase in total population occurred in Beijing's central area, which constitutes only 8.34% of the city's total land area. The downtown population density was, on average, 8972 people/km^2, whereas the suburban areas had a population density of 559 people/km^2 [2]. Without a comparably developed suburban area to absorb the population growth, the population concentrated in the central area. That condition resulted in Beijing's development of the special passenger transportation system and major traffic overloads.

2.2. The development of demand for passenger transit

The demand for passenger transit in Beijing has soared in response to rapid economic development, the expansion of the population, and short-sighted urban planning with intensive land use for a single purpose. In 2013, the total number of daily trips (not including trips on foot) was 30.99 million in the city's urban centre, an increase of 12.67 million at the end of 2003. The growth in the number of trips significantly exceeded population growth. In the central area, the average number of trips per person increased from 1.25 in 2003 to 1.47 in 2013. The increase in the average number of trips can be ascribed to the increasing prosperity of Beijing's urban economic activities and to single-purpose land use. Large areas of land have been dedicated as residential or other functional zones and residents are compelled to travel to satisfy their various needs (e.g. shopping). This situation is particularly evident regarding travel from extremely large residential areas to central business districts.

2.3. The development of traffic supply

Various transportation infrastructures and transit measures were created in Beijing to attempt to meet the soaring traffic load. Regarding public (mass) transit, at the end of 2013, there were 813 bus lines with 23,592 buses covering 19,688 km. There were 17 metro lines serving 465 km with 3998 operating metro vehicles [3]. For comparison, in 2003, there were 6839 fewer buses (a total of 16,753 buses) and 3306 fewer operating metro vehicles (a total of 692 metro vehicles) in service. Travel by taxi in Beijing is regarded as a high-comfort mode that is preferred by high-income passengers. Therefore, its use is relatively low and the taxi fleet is strictly controlled. Although the number of passengers that travel by taxi is assumed to be steadily increasing, there has been relatively little change in the number of taxis, which increased from about 62,000 in 2003 to about 67,000 in 2013.

2.4. Passenger transportation mode-share

The evolutionary trend of Beijing's urban passenger transportation mode-share proportions between 2005 and 2013 suggests that public transit's share steadily increased and private passenger road vehicles' share was effectively controlled (likely by the enforcement of legislative regulations), as shown in Figure 1. With regard to the transportation modes in 2013, rail transit's share was about 20.60%, an increase of 3.80% over 2012. Buses' share was about 25.40%, a decrease of about 1.80% and private passenger road vehicles' share was about 32.70% of the total trips. Private passenger road vehicles' share increased only 0.10%, whereas taxis' trip share decreased by 0.20% to 6.50%. Bicycle's share increased by 12.10%.

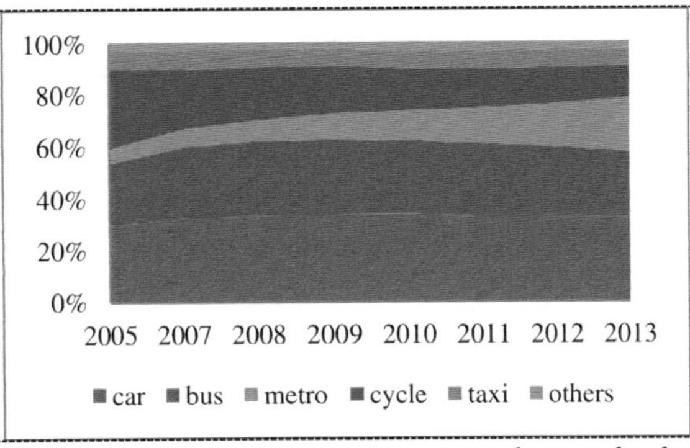

Figure 1. Evolutionary trends in passenger transportation mode-shares, 2005–2013.

3. Passenger transportation policies

Passenger transportation polices can influence the transportation demand as well as the supply and, thus, play a significant role in the evolution of passenger transportation. The history of the evolution of metropolitan areas in the world demonstrates that logical and coherent transportation policies can guide consumers' purchases and uses of private passenger road vehicles, optimize passenger transportation mode-sharing, and promote mode-shifts to modes with lower carbon emissions. For example, Beijing has implemented a series of transportation policies to emphasise public transit and regulate private passenger road vehicle purchases and use.

3.1. Private road vehicle ownership and use regulations

Passenger transportation regulations were temporarily implemented before the Beijing Olympic Games (July through September 2008) to reduce the number of private passenger road vehicles on the roads and improve air quality. After the Games concluded, in October 2008, Beijing changed the policy into a formal and permanent regulation. The new version forbids a private passenger road vehicle to be used on one day per work week (Monday through Friday) between the hours of 7:00 and 20:00 (CST). The restrictions were intended to partially mitigate Beijing's traffic congestion and they were successful in that the reduction of private passenger road vehicles in the city reduced the congestion index from 7.70 in 2007 to 5.40 in 2009. However, the improvement lasted only two years because the total number of motor vehicles continued to increase and the effects of use regulations were offset by increases in road vehicle ownership.

Passenger road vehicle ownerships soared from 2.12 million vehicles in 2003 to 4.81 million vehicles in 2010. The congestion index rebounded to 6.10. In response, the Beijing municipal government introduced several measures in December 2010 intended to alleviate traffic congestion, including a lottery registration for road vehicle purchases, increased parking fees in the downtown area, and accelerated construction of the public transit system. Some large cities in developed countries set limits on private passenger road vehicle ownership and have vehicle use policies to combat traffic congestion.

For example, in February 2003, London began collecting a fee (congestion charge) from private passenger road vehicles entering a section of the city centre. The effects of that policy are apparent in a report by Transport for London indicating that the number of private passenger road and minicab vehicles entering the 'charge zone' decreased by 30%, the number of vans and lorries decreased by 13%, and the number of trips on buses and, particularly, on bicycles increased [4]. A congestion charge is likely to be implemented in Beijing.

In 2011, the regulations limiting private passenger road vehicle ownership were implemented in Beijing. The purchase quota per month of 10,000 vehicles was allocated in a license plate lottery. The result has been a dramatic decline in the growth rate of private passenger road vehicle ownership. In 2013, the annual growth was 4.56%, much lower than the 19.70% growth rate in 2010. Under the pre-2011 conditions (without controls), private passenger road vehicle ownership in 2013 would exceed six million.

Despite this success, traffic congestion and pollution problems differ by region and different approaches should therefore be taken to account for those differences. A blanket policy across the country would likely be inconsistent in its effectiveness. Regulations on ownership are likely effective and should be implemented in central and high-population density areas to counter rapidly increasing private passenger road vehicle ownership. However, in suburban areas where public transit service is not very effective and private passenger road vehicles may be a necessary supplement, regulating ownership may cause economic hardship. Thus, big cities should devise flexible policies regarding the technical, economic, legislative, and administrative aspects of transportation management. The current blanket policy cannot accommodate the diversity in demand for number and length of trips in different areas and at different times. In fact, almost one-half (47.8%) of drivers sometimes ignore the use control rules [5]. Specifically, those drivers whose destinations are far from the city centre or metro stations and who travel during peak hours are relatively more likely to violate the use control rules.

3.2. Public transit incentives

Beijing has long regarded its public transit system as a public welfare infrastructure. Government subsidies have historically resulted in bus and metro fares that were significantly lower than in other major Chinese cities. Fares were further reduced in 2007 so that buses had a base fare of CNY 1 and passengers who hold the IC card (a discount card that is swiped to pay the fare) enjoyed a 60% discount. In addition, the student card provided an 80% discount. This policy was intended to attract travellers to the bus or metro to promote public transit and make public transit more affordable to low-income people. With the stimulus of lower fares, metro ridership increased from 654.93 million in 2007 to 1.22 billion in 2008. However, there were negative unintended consequences. First, although the 2007 policies attracted more passengers, public transit's overcrowding problem worsened and the density of passengers standing in metro cars exceeded seven persons/m^2 [6]. Second, the extraordinarily low fares triggered invalid demand that resulted in ever-increasing governmental subsidies of public transit operations.

To address the crowding problem and tight budget, a new fare policy for the bus and metro systems was launched in December of 2014. A flat fare of CNY 2 for metro use was replaced with a distance-based fare structure with a base fare of CNY 3 for trips within a six-kilometre radius. The base fare for bus trips was increased to CNY 2 for trips within a 10-kilometre radius plus CNY 1 charged for each additional 5 km travelled. However, bus fares were 50% discounted for IC cardholders and the student card discount was 75%.

3.3. Public transit subsidy

The Beijing government provides large subsidies to public transit because it is state-owned, fares are kept low, and the public transit system is growing by adding new bus and rail lines. In 2009, the public transit subsidy was about CNY 12.93 billion; four years later in 2013, it had increased to CNY 20.01 billion. Increasing subsidies created persistent financial pressures and the Beijing government launched a reform of its investments in the public transit infrastructure by integrating private financing through a public–private partnership scheme. The reform is expected to mitigate the financial pressures on the government, boost the efficiency of transit operations, and improve the desirability of public transit by increasing the level of services.

4. Current issues

Although Beijing's passenger transportation complex is evolving in a positive direction, many of the transportation management policies have achieved initial success, and the mode-share of private road vehicles appears to be at a historical stagnation or may even be in a declining trend, salient problems threaten the sustainability of Beijing's passenger transportation systems.

4.1. Aspects of passenger transportation needing improvement

First, the mode-share of public transit still has potential for improvement and road traffic congestion and public transit crowding problems are still serious. Public transit's and bicycling's mode-shares are relatively low and residents are relying less and less on bicycles. Second, private passenger road vehicles are still the main travel mode not only because ownership is large; their use is not effectively regulated. The average daily number of trips is 1.86, which is much higher than in Tokyo or Paris. Road congestion is a major problem in Beijing's central area and the average travel speed of private road vehicles is low. Third, the urban public transportation systems have some urgent issues. For example, public transportation congestion seriously influences mode-shifts during a trip. Currently, Beijing's public transportation has an in-vehicle (bus or metro) passenger density during the morning peak hours that exceeds 7 people/m^2 [6]. Suburban railways, which have an important role in dispersing the central area populations to the suburbs by providing convenient long-distant commuter travel options in cities such as London and Tokyo, have not been well developed in Beijing.

4.2. Need for planning and design integration

The planning and design of changes made to the different transportation modes are not integrated, which results in inconvenience to the riders when they make transfers between modes. At present, the rail network and the bus line network cannot achieve a high level of mutually beneficial operations due to the inconvenient interchanges and the competing demands between bus lines and metro lines that service the same corridors. The distance between feeder bus stations and the corresponding metro stations is generally too great. Moreover, confusion is added to the problem because of unclear signage. Suburban park–and–ride facilities, which are crucial for transitioning the public from private road vehicles to public transit, are not always included in land-use planning.

4.3. Budgetary problems

The low-fare policy and inefficiency of the station-owned public transit operation companies create significant pressures on Beijing's fiscal budget. In addition, the total dynamic investment in the urban rail transit construction expected before 2035 was estimated as CNY 627.70 billion [7]. A variety of investment modes were attempted, but only CNY 23 billion was obtained from the private sector between 2000 and 2014 [7]. This shortfall is worsened by the financial problems of the bus and metro lines' operating companies in covering current operational costs with revenues, which fall far short, requiring more than CNY 20 billion annually in subsidies. The deteriorating financial situation of the public transportation infrastructure is likely primarily attributed to the low public transit fares. However, the inefficiencies of the state-owned corporations also contribute. The situation is complex: The government cannot distinguish one influence from the other without effective financial assessing indicators and subsidies are provided based on the financial statements of the operation companies.

4.4. The underlying fundamental problem of population density

The problem of overwhelming trip demands in space (land area) and time (peak hours) is generated by high-density and single-function land use. This is the fundamental issue and it is seldom considered in Beijing's urban planning. The extremely large residential areas that have been developed, such as

Tiantongyuan and Huilongguan, create very uneven traffic demands across time and the current transportation options in these areas are insufficient at peak hours yet exorbitant during off-peak hours. Although this problem is obvious, planners apparently did not learn from the mistakes. About 50% of the central business district (or the equivalent in trip generation districts) was developed in the existing centralized area of Beijing during the past five years, which almost completely offsets the benefits derived from the rapid development of the metro and bus systems for transporting passengers to outlying suburban areas.

4.5. Air pollution

Urban passenger transportation is the primary cause of Beijing's air pollution. In 2013, motor vehicles (of all classes) emitted about 22.20% PM2.5 (particulate matter), 58% NO_2 (nitrogen oxide), and 40% VOCs (volatile organic compounds) [8]. During the Asia-Pacific Economic Cooperation (APEC) Conference in Beijing, strict restrictions on road vehicle use were implemented and large institutions in central Beijing were given short vacations to reduce the amount of traffic. The 'APEC blue sky' seen during the event was evidence that private passenger road vehicles in Beijing are a crucial element of environment problems.

5. Conclusions and suggestions

This study employs Beijing as a case study from which to derive suggestions for general application in late-developing cities. Based on Beijing's patterns of passenger transportation development over the past decade, transportation mode-sharing was evaluated, causes and effects of implemented transportation policies were analysed, and the persistent problems were described and summarized. Based on the conclusions drawn from the analysis and discussion, planning strategies and suggestions for the development of passenger transportation systems in big cities in developing countries emerged. First, urban spatial growth and urban population increases are the inevitable outcomes of emerging economies. As residents' incomes increase, private passenger road vehicle ownerships and the extent of use will increase. This is the most important issue for urban passenger transportation. To address the problems that necessarily occur as a result, cities must take a two-pronged approach. They must improve and expand their public transportation systems as the population grows and they must regulate the ownership and use of private passenger road vehicles. In addition, regulations should be implemented before serious problems arise as opposed to after the fact and regulations should be flexible enough to be effective in diverse contexts. Beijing's regulations were too late and too rigid to be thoroughly effective. The blanket policy was unnecessary on low-traffic roads in the suburbs. Nevertheless, the regulations achieved positive results and the share of trips by private passenger road vehicles has been decreasing. The Beijing government is now assessing ways to enforce a congestion charge on private passenger road vehicles used in the central area to further cut their mode-share as well as ways to increase the flexibility of regulations placed on use.

Second, developing the transportation infrastructure and improving the service level of the public transit systems are essential to transition the public from reliance on private passenger road vehicles to public transit. Convenient and comfortable alternatives to private passenger road vehicles should be offered. The large-capacity public transportation systems should be oriented at the corridor-level mode. In the case of Beijing, the metro system effectively contributes to the mode-shift. However, so long as the level of service provided by the metro suffers from the overcrowding problem, residents will tend to use private passenger road vehicles or taxis. If the long-term rail transit plan of 1000 km lines, which includes metro, light rail, and suburban railway, is implemented, the public transportation service level may dramatically improve and thereby attain the dominant mode-share.

Third, the primary challenge to providing more and better public transit is financing. At present, the average annual investment in metro construction in Beijing is about CNY 20 billion, of which almost 60% is bank loans [7]. The lack of money creates a great deal of pressure on the government's future financial allocations. Moreover, to maintain the low fare policy, the government must provide subsidies of more than CNY 20 billion annually to public transit. Therefore, cities in developing countries with tight budgets must reform their investments in and financing of their public transportation infrastructure construction needs and focus as much as possible on private capital. One possible way to attract private funding is by adopting new financing modes, such as franchises.

Furthermore, to reduce high operating subsidies, a reasonable transit fare structure should be established based on daily transportation expenditures and residents' or users' incomes and abandon unrealistically low fares. To control operating costs of and subsidies to public transit, competitive bidding among operating companies and effective operating assessment schemes should be adopted. Beijing has implemented investments and financing modes such as BT, PPP, and sale-leaseback in the metro system. This action raised capital for public transportation infrastructure construction and introduced new metro operation companies, which created a competitive relationship among state-owned enterprises and private companies that should drive down costs. Although these measures are at their initial phases, the level of service has improved and operating subsidies are lower. In the future, these measures should be more widely and extensively implemented.

Urban traffic problems in large cities primarily result from rapid economic and population growth and the corresponding land development. In the case of Beijing, the over-concentration of development in the central area, particularly large-scale residential development in the downtown area, has generated a high and uneven demand for passenger transportation. The only way to meet this need is to increase the supply of transportation options on a grand scale, which often requires major investments that have limited effects. Therefore, emerging large cities in developing countries must integrate their urban planning with the development of their passenger transportation systems. Land-use planning should be based on the relationship between land use and traffic. By doing so, passenger transportation concerns can be pre-emptively handled before they become major problems for the government, the public, and the environment.

Acknowledgments
This work was financially supported by the Project funded by central non-profit research institute basic research funding (Project No. 20194807) and Fujian Provincial Transportation Administration Project "Research of collaborative linkage technology of Urban transportation subway and bus system synergy linkage technology"(Project No. [3500]SG[GK]2019003). Beijing Infrastructure Investment Corporation Limited, Beijing Subway Operation Corporation Limited and Beijing MTR Corporation Limited were thanked for providing raw data, project documents and interview.

References

[1] Beijing traffic development research center (2014) Beijing traffic development annual report in 2013(in Chinese).
[2] Beijing municipal bureau of statistics (2014) Beijing national economic and social development statistical bulletin in 2013 (in Chinese).
[3] Beijing municipal bureau of statistics (2014). The situation of Beijing in 2013(in Chinese). Available on: http://www.bjstats.gov.cn/bjsq/csjs/
[4] Transport for London. (2007) Travel in London Report 6.
[5] WANG, L., XU, J. & QIN, P. (2014) Will a driving restriction policy reduce car trips? The case study of Beijing, China. Transportation Research Part A: Policy and Practice, 67, 279-290.
[6] CHEN, F., WU, Q., ZHANG, H., LI, S. & ZHAO, L. (2009) Relationship Analysis on Station Capacity and Passenger Flow: A Case of Beijing Subway Line 1. Journal of Transportation Systems Engineering and Information Technology, 9, 93-98.
[7] Zhang G., Liu, Y., Xu, X., and Chen, F. (2012) Innovation of investment and financing theory and practice of Beijing's urban rail transit. Tsing Hua University Press, Beijing (in Chinese).
[8] Environment Protection Bureau of Beijing Municipal Government.(2013) Beijing's Air Quality Management Initiatives(2013-2017). Available on: http://www.kankyo.metro.tokyo.jp/policy_others/international/attachement/Session1.pdf

Analysis and Thinking of Urban Transportation Planning against the Backdrop of Multi-planning Integration

Yixiao Yin[1*], Zhong Guo[1] and Yuanwu Tang[2]

[1]China Urban Sustainable Transport Research Center, China Academy of Transportation Sciences, Beijing, 100029, China

[2]Jangcun railway station, China Railway Guangzhou Group Co.,Ltd., Guangzhou, 510450, China

[*]violayin730@qq.com

Abstract. In recent years, China has shifted from high-speed growth to high-quality development. At the same time, its urban transportation system is also moving from rapid expansion to intensive and efficient development. Though urban transportation system has entered the stage of relative maturity and stability, some problems are impacting heavily on plan implementation, e.g., poorly-developed planning system, overlapped planning management, theory & practice decoupling and lack of some evaluation steps. In the new situation of multi-planning integration, we propose integrating urban transportation planning into the spatial planning system by focusing on transforming and upgrading transportation planning management, and by streamlining the planning relations, unifying planning system, coordinating planning management, regulating planning compilation and improving implementation mechanism, so that urban transportation planning plays the greater role of strategic orientation in multi-planning system.

1. Introduction

In the traditional planning model in China, plans for the national economy and social development, urban and rural development, land use, ecology and environment, and transportation are made separately, thus lacking coordination. Because plans differ in focal points, targets, technical standards, methods and means of compilation, and because overlapped and even conflicting plans are often seen[1]. The lack of coordination among plans has seriously impacted on efficiency of plans nowadays when resources and environment are facing great pressure.

In recent years, the CPC Central Committee and the State Council of China have emphasized on multi-planning integration, requiring for many times facilitation of multi-planning integration, building spatial planning system, speeding up development of ecology, and rationally allocating natural resources. At the same time, when the Chinese cities are evolving from high-speed expansion to high-quality development, external environment of urban transportation has changed, highlighting resource and ecological constraints, triggering changes in demand-supply characteristics of urban transportation system, and producing evident demands for high-quality service. In light of these factors, the paper studies tactics of urban transportation planning in the background of multi-planning integration, enhances coordination between urban transportation planning and multi-planning in terms of compilation and management, integrates urban transportation planning into the system of spatial planning, ensures the urban transportation planning guides urban space structure and land-use pattern, implements the philosophy of "transportation steering urban development" and realizes intensive, orderly and high-quality urban development.

2. Current development of urban transportation planning

Up to date, China has not issued a special law for urban transportation planning. The Law of Urban and Rural Planning of the People's Republic of China requires compiling urban transportation plan as a special part of urban plans. As the non-statutory plan, urban transportation planners should follow such laws and regulations as the Law of Urban and Rural Planning of the People's Republic of China, the Law of Compiling Urban Plans and incumbent urban policies. They are guided by relevant technical specifications such as Specifications of Urban Road Transport Design and Plan, Methods of Making Urban Transportation System Plans and Guidelines for Making Urban Transportation System Plans. At the same time, the urban transportation plan should stay in line with major national-level plans of infrastructures, regional plans, urban master plans, and urban land-use plans, etc.

2.1. Planning system

Current guiding policies in China mainly specify contents and requirements for urban transportation plans, without regulating the urban transportation planning system. Some cities, based on their own characteristics, study urban transportation planning systems. In general, urban transportation planning system includes two stages and three levels. Two stages refer to the stage of master plan and stage of special plans. Three levels refer to urban transportation development strategic plan, urban long and mid-term transportation system plan and urban transportation recent treatment plan[2]. The urban transportation development strategic plan and urban long and mid-term transportation system plan mainly correspond to urban master plan. However, in practice, the Chinese cities, when making urban transportation plans, try to make a holistic plan to cover all technical contents. The plan that attempts to include all often results in sparse targets, poor revolution of major problems, superficial work and lack of strategic vision.

2.2. Problems and cruxes

After years of practice, urban transportation planning has entered the stage of relative maturity and stability. However, there are still some problems and limitations in the process of making and carrying out these plans, which make it impossible to achieve the expected results when carrying out the plans. First, fail to fully recognize the position and role of the plans, thus resulting in low rate of compilation. Urban transportation plans are often used to replace special transportation plans and transportation improvement plans. Most of the small and medium-sized cities don't have plans. Second, planning philosophy is decoupled from practice. Planning philosophy emphasizes on people while plans focus on vehicles; though the plans claim to adopt the strategy of public transport priority, measures are not seen in these plans; third, evaluation of plan implementation is not included in the regulated procedure. There are no specific deadlines or content requirements for evaluation. As a result, the plans lack effectiveness and continuity[3].

The crux of the problems are:

- Statutory position lacking. Transportation plans lacking protection of laws, which made their position is much lower than other statutory plans. It is impossible for transportation plans to stay equal to other plans in the multi-planning integration;
- Incomplete planning system and backward planning theories and methods. Urban transportation planning function and relations were ambiguous, and the formulation of planning goals lacks scientific evidence and is often difficult to achieve;
- Separation and overlapping of functions and powers of the existing administrative system. The lack of coordination and management mechanism in urban traffic planning and management affects the connection of urban traffic planning, implementation and management.

3. Requirements of multi-planning integration for urban transportation planning

3.1. Connotations of multi-planning integration

The multi-planning integration refers to connection and coordination of plans for national economy and social development, land use, urban and rural development, ecology and environmental protection, transportation, education and public health, etc. It is an important measure to establish spatial planning system and facilitate establishing regulations for ecology. Within their own power scope, local governments coordinate the various spatial plans to make sure the multi-planning integration has the same spatial parameters in protective space, development borders and city scale, etc. At the same time, on the spatial information platform, they establish control line system, optimize spatial layout, effectively allocate land resources and improve spatial control[4].

3.2. Relevance of urban transportation planning with multi-planning integration

In China's system of spatial plans, what are related to urban transportation plans are national economy and social development plan (economy plan), land use master plan (land plan), urban master plan (master plan) and environmental protection plan (EP plan). As the upper planning basis, they mainly guide and constrain urban transportation planning and construction at the macro level. Urban transportation planning is often in a passive position to interact and connect with the master plan though they are made simultaneously. There is almost no "reverse feedback" with "economy plan", "land plan" and "EP plan".

3.3. Requirements of multi-planning integration for urban transportation planning

Though relevant laws and regulations raise requirements for coordination and connection of plans, they do not specifically regulate how and where the plans get connected. As a result, urban transportation plans are other plans have the following problems: planning laws and regulations differ greatly in system; the plans differ greatly in position; plan management mechanisms are not balanced; planning periods are not balanced; and technical standards are not the same, etc. These problems have great impact on integrating the urban transportation plan into the other plans.

In the situation of multi-planning integration, to facilitate integration of urban transportation plans into spatial planning system, solve current planning problems and improve effect of carrying out the plans, the following requirements are raised for urban transportation planning.

- Unifying the planning system, clarifying the role of plans in various levels, abide by and support urban structure and functions determined by spatial strategic plans, abide by requirements of control line, and handle properly the connection of transportation plans with other plans;
- Coordinating plan management, regulating plan making, coordinating resource allocation and optimization among various systems in urban transportation, unifying the planning periods and basic data processing, and handling duty division and transfer of transportation plans in various levels;
- Establishing mechanism that secures implementation of plans, clarifying duty division and coordination measures to make, carry out and manage the plans. For overlapped contents in both transportation plans and space control of other departments, it is necessary to coordinate them in the guidance of spatial planning.

4. Experience to learn

4.1. Establish a coordinating agency for planning

The laws of the United States required cities with more than 50,000 population to set up MPO, which is the transportation decision-making organization consisting of representatives from local government and transportation groups[5]. MPO is responsible for coordinating and solving transportation development problems in the region, including land use, air quality, energy, economic growth and commerce, etc. By establishing coordinating agency in the region, US establishes mechanism of inter-department negotiations to provide a relatively effective joint decision-making platform for regional cooperation, breaking the borders among administrative zoning and departmental duties for regional transportation planning, and realizing the multi-planning integrity of plans for economy and social development, land use, population, transportation and environment, etc. in the region.

4.2. Establish management system of "greater transportation"

By making "greater transportation reform", Japan has established the Ministry of Land, Infrastructure, Transport and Tourism, including the previous Ministry of Infrastructure and Ministry of Land into the core department of greater transportation in the country. The Ministry supervises various means of transportation such as highways, waterways, railways and flights. At the same time, it is responsible for construction, state land and tourism, etc[6]. The Ministry of Land, Infrastructure, Transport and Tourism consists of the headquarters and external bureaus. The headquarters are composed of internal bureaus and departments, special agencies and local branch bureaus. The external bureaus include Seamen Labor Committee, Japan Coast Guard, Japan High Marine Accidents Inquiry Agency and Japan Meteorological Agency, etc. In terms of department and duty setup, Japan ensures hierarchy management, clarifies hierarchy management of central government and local governments, secures the uniform coordination of the central government and fully mobilizes initiative of local administrations.

4.3. Planning system with clear-cut layers

The developed urban transportation systems in Beijing and Shenzhen are build and operated in an efficient way due to the relatively well-developed urban transportation planning system, which ensures the sub-systems of urban transportation are built and operated in a rational and fact-based way.

Beijing has set up a four-layer urban transportation planning system, including strategic planning, comprehensive planning, regional planning and special planning, and well corresponding with each stage of the urban master plan[2]. While Shenzhen has gradually established four-level and two-stage urban transportation planning system. Four levels includes holistic planning, detailed planning, comprehensive improvement planning and annual evaluation implementation, and two stages are transportation planning (TP) and transportation improvement planning (TIP), covering the transportation solutions in the whole process[7].

5. Proposals on urban transportation planning in the background of multi-planning integrity

The integration of urban transportation planning and multi-planning is both a technical problem and a institutional issue. In summary, the following proposals are put forward for promoting better integration of urban transportation planning into the spatial planning system:

5.1. Unify planning system and form synergy

Speed up building the urban transportation planning system that has "accurate positioning, clear borders, complementary functions and connection". Study and formulate the Framework of Urban Transportation Planning System to standardize the urban transportation planning system. Learn to divide the planning levels and contents, keep coordination with the urban planning system, and establish the urban transportation planning system covering four levels of "strategic planning, comprehensive planning, special planning and implementation planning". According to the framework, each city refines specify planning system suitable for the needs. Improve the top-level design of the plan, and strengthen the coordination between the plans. Horizontally, the structure and working phases of the urban transportation planning system are consistent with urban planning system, and the planning scope, base map, and period are unified. Vertically, adhere to the coordination between subordinate planning and superior planning, and clarify the planning framework, content, depth requirements and key connection points of each level.

5.2. Clarify positioning and rationalize relations of planning

Based on the requirements for the multi-planning integrity, clarify the functional positioning of urban transportation planning, rationalize the interrelationships between urban transportation planning, take the transportation development strategic planning as the guidance, the comprehensive transportation planning as the basis, and the special planning and implementation planning as support to avoid cross duplication and conflict. The strategic planning determines the situation of the urban transportation system at the macro-strategic level, working at the guideline of all other levels of planning. The in-depth study of comprehensive transportation planning based on strategies, which is the basis for the government to coordinate and regulate transportation resources, support urban social and economic development, and guide various special planning. The special planning is detailed to implement the strategic tasks of the integrated transportation plan for the subsystems. The implementation planning guide the recent transportation construction, construction and renovation, impact assessment, and engineering design.

5.3. Coordinate planning management and regulate planning production

On the one hand, we should improve the institutional mechanisms for the management of urban transportation planning. First, adhere to the combination of comprehensive coordination and departmental linkage, establish a joint conference system and consultation and communication mechanism involving natural resources, development and reform, housing construction, environmental protection, transportation and other departments, and clarify all parties' duties. Second, determine the legal procedures for public participation, expand the scope of comments, and increase public participation in planning. Besides, we accelerate the establishment of a planning management information platform to ensure the unification of basic data and base maps for urban transportation planning and other spatial planning. On the other hand, we will accelerate the development of national-level urban transport planning laws, regulations, and standard systems. Accelerate the formulation of the "Law of the People's Republic of China on Urban Transport Planning" to ensure the legal status. Revise and improve the existing standards and guidelines for urban transportation planning, clarify and standardize the preparation procedures. Use Internet technology to promote innovation in transportation planning technology and methods.

5.4. Improve mechanism of implementation to secure plans is carried out

In accordance with the principle of “those spearheading the production of plans are responsible for implementing them”, we should strengthen the evaluation of planning implementation, clarify and implement the responsibilities of the entities, establish a dynamic adjustment mechanism, improve monitoring and evaluation of planning implementation, and improve the effectiveness of planning implementation.

First, we propose strengthening planning and implementation evaluation. Adopt the “city government-led, transportation planning + action plan decomposition and implementation” model, incorporate planning implementation evaluation into prescribed procedures, clarify planning implementation evaluation requirements, and quantify planning evaluation results. Second, establish a dynamic planning adjustment mechanism. According to the results of the planning evaluation, set the conditions for planning adjustment and revision, establish a dynamic update system for the planning “making-implementation-evaluation-adjustment”, and set a fixed period for planning evaluation and updating. At the same time, we should improve the planning and implementation of the supervision and evaluation mechanism. The planning and preparation department shall make planning and implementation as an important part of the disclosure of government affairs information, consciously supervise the people's government and the public, and explore the implementation of a mechanism for linking the results of the assessment of planning implementation with the performance of the subject responsible for the assessment.

6. Conclusion

In the new situation of the new era, the development of urban transportation planning should take a long-term perspective and improve the top-level design. We should pay attention to innovation of the policy mechanism and improvement of technological argumentation, promoting the urban traffic planning to be integrated into the spatial planning system, enhancing the integration of the traffic planning with different resources and realizing coordinated development of traffic with urban development, land utilization and environmental protection.

Acknowledgments

The research was financially supported by the Ministry of Transport Projects "Research on Urban Comprehensive Transportation Planning under the Background of Multi-Regulation Integration" (No.2017-14-6).

References

[1] Yingting, X., Wei, W. (2015). From "multi-plan integration" to the reconstruction of spatial planning system. Urban Planning Forum, 3: 15-21.
[2] Yao, S., Zhang, Z., Kong, L., Xiong, Z. (2016). Beijing Urban Traffic Planning System. Transport Research, 2: 38-44.
[3] Wang, J. (2017). Review and Prospect of Urban Comprehensive Transportation Planning in China. Urban Transportation of China, 15: 18-24.
[4] Wang, G. (2018). Urban Multiple Plan Integration and Urban Transportation Planning Reformation. Urban Planning Forum, 5:19-28.
[5] Shen, S., Wang, X., Liu, Z. (2016) Inspirations of the Regional Transportation Planning System of MPOs to Chinese Syncretic Planning. Comprehensive transportation, 10:85-93.
[6] Lu, X. (2011).Japan's Land Planning Reform Promotes Urbanization Process and Its Enlightenment to China. Urban Studies, 5: 34-37.
[7] Zhang, X., (2016). Innovation and practice of urban traffic planning in Shenzhen. In: 2016 China Urban Transportation Planning Conference. Shenzhen. 2016:838-847.

IOP Conf. Series: Materials Science and Engineering **792** (2020) 012003 doi:10.1088/1757-899X/792/1/012003

Application of grey system theory in prediction of accident prone sea area

Wang xi-kun[1], Wang yi-xin[1] and Xu you-hao[1]

[1] Shandong Jiaotong University,Weihai, Shandong, 264200, China

*Corresponding author's e-mail: 2564546101@QQ.com

Abstract. The GM (1,1) model is established by using the grey system theory,Realizes the goal that the accidents in the multi accident sea area can be accurately predicted by the prediction model. The key point is to establish the model correctly and apply it to the prediction of accident prone sea area. Looking up the sea area with the characteristic of frequent accidents, building the model, and testing the result of substituting the data into the model prediction, the result shows that the viewpoint of this paper is correct and can predict the maritime traffic accidents with high accuracy.

1. Introduction

GM (1,1) model of grey system is based on the comprehensive data of many known factors in the system, and the time series of this data is fitted by differential equation to force the dynamic process described by "near time series", and then extrapolate to achieve the purpose of prediction.

2. Grey system theory and GM (1,1) model

2.1. On grey system theory and GM (1,1) model

The original sequence is recorded as:

$$x^{(0)}=\{x^{(0)}(1)\ ,\ x^{(0)}(2)\ldots\ldots x^{(0)}(t)\} \quad (2.1)$$

Due to the randomness of the occurrence of ship accidents, there is no specific law. To weaken the randomness of ship accidents, we need to adopt the method of accumulation to deal with the time data and get the accumulated generated number as follows:

$$x^{(1)}(t)=\sum_{i=1}^{k}x^{(0)}(i)\ ,\ldots\ldots,\ n \quad (2.2)$$

Thus, we can get the generating terms without less randomness and volatility. The original sequence of the cumulative generating number is recorded as follows:

$$x^{(1)}(t)=\{x^{(1)}(1)\ ,\ x^{(1)}(2)\ ,\ kx^{(1)}(k)\},\ k=1,2,\ldots\ldots,\ n \quad (2.3)$$

The number of progressive generation is:

$$x^{(0)}(k)=x^{(1)}(k)-x^{(1)}(k-1)\ ,\ x^{(1)}(0)=0. \quad (2.4)$$

Then, according to a series of time data obtained above, $x^{(1)}$ change shows exponential change, and the new data sequence is assumed to satisfy the first-order univariate ordinary differential equation:

$$\frac{dx^{(1)}(t)}{dt}+ax^{(1)}(t)=b \quad \text{(a and b are constants not equal to 0)} \quad (2.5)$$

IOP Conf. Series: Materials Science and Engineering **792** (2020) 012003 doi:10.1088/1757-899X/792/1/012003

$$\begin{bmatrix} a \\ b \end{bmatrix} = \left(B^T B\right)^{-1} B^T Y_n, \quad D^{(1)}(k+1) = -\frac{1}{2}[x^{(1)}(k) + x^{(1)}(k+1)] \quad (2.6)$$

Formula:

$$B = \begin{bmatrix} D(1) & 1 \\ D(2) & 1 \\ \vdots & \vdots \\ D(n) & 1 \end{bmatrix}, \quad Y_n = \begin{bmatrix} x^{(0)}(2) \\ x^{(0)}(3) \\ \vdots \\ x^{(0)}(n) \end{bmatrix} \quad (2.7)$$

The prediction model is established by data accumulation:

$$x^{(1)}(k+1) = [x^{(0)}(1) - \frac{b}{a}]e^{-ak} + \frac{b}{a} \quad (k=0, 1, 2, \ldots, n) \quad (2.8)$$

The final grey prediction model is as follows:

$$\bar{x}^{(0)}(k+1) = \bar{x}^{(1)}(k+1) - \bar{x}^{(1)}(k) \quad (2.9)$$

2.2. *Test the model*

First, the model is tested by residual and relative residual methods:

$$E(t) = x^{(0)}(t) - x^{(-0)}(t) \quad (2.10)$$

$$e(t) = \frac{E(t)}{x^{(0)}(t)} = \frac{x^{(0)}(t) - x^{-(0)}(t)}{x^{(0)}(t)} \quad (2.11)$$

In two formulas $x^{(0)}(t)$ is the number of ship traffic accidents in the original year t, $x^{-(0)}(t)$ is predict the number of ship traffic accidents that will occur in the T year, $e(t)$ the smaller the model is, the higher the prediction accuracy is; otherwise, the opposite is true.

In addition to the above methods, the posteriori difference test is also a conventional test method.

Posterior error test: set S1 as the mean square deviation of the original ship traffic accident sequence and S2 as the mean square deviation of the above residual sequence, and calculate them separately.

$$S1^2 = \frac{1}{n}\sum_{k=1}^{n}[x^{(0)}(k) - \bar{x}^{(0)}]^2 \quad (2.12)$$

$$S2^2 = \frac{1}{n}\sum_{k=1}^{n}[E(k) - E]^2 \quad (2.13)$$

$C = S2/S1$ and $P = P\{|E(t) - E| < 0.6745 S1\}$ the ratio of posterior error and the frequency of small error, respectively, is an index to measure the accuracy of the prediction model. The smaller the C is, the higher the accuracy is, the larger the P is, the higher the accuracy is, and vice versa. The combination of these two indicators can comprehensively determine the accuracy of the prediction model. Generally speaking, the accuracy of the model can be divided into four grades (see Table 1).

Table 1 Accuracy class

Accuracy grade	e	k_0	C_0	p_0
LEVEL 1	0.01	0.90	0.35	0.95
LEVEL 2	0.05	0.80	0.50	0.80
LEVEL 3	0.10	0.70	0.65	0.70
LEVEL 4	0.20	0.60	0.80	0.60

3. The practice of ship traffic accidents in multi accident sea area in grey system theory

3.1. Recent accidents in Liaoning and Qingdao sea areas

Table 2 and table 3 respectively show the number of maritime vessel traffic accidents in Liaoning sea area and Qingdao sea area, which are frequent accidents in China from 2012 to 2018.

Based on the investigation of ship traffic accidents in recent years, the data in the following table are obtained:

Table 2 Traffic accident volume of ships in Liaoning sea area in 2012-2018

Particular year	2012	2013	2014	2015	2016	2017	2018
Traffic accident volume	165	159	154	149	148	150	148

Table 3 Traffic accident volume of ships in Qingdao sea area of China in 2012-2018

Particular year	2012	2013	2014	2015	2016	2017	2018
Traffic accident volume	154	149	142	137	135	138	136

The data of ship traffic accidents in these two accident prone areas are superposed and integrated into a new table 4.

Table 4 Total traffic accidents in Liaoning sea area and Qingdao sea area

Particular year	2012	2013	2014	2015	2016	2017	2018
Total traffic accidents	319	308	296	286	283	288	284

Through table 4, we can directly read the total number of ship traffic accidents in the multi accident sea area from 2012 to 2018.

3.2. Modeling

First, we directly use the original data of ship traffic accidents to build the model, accumulate the original data of ship traffic accidents from 2012 to 2018, and get the original sequence:

$$[x^{(n)}/432,397,358,326,298,312,281] \quad (3.1)$$

Finally, the GM (1,1) prediction model is obtained as follows:

$$x^{-(1)}(t+1) = -4127.215139e^{-0.109362*k} + 5174.216951 \quad (3.2)$$

Test the model: $\bar{e} = 0.12$, $e = 0.11$,test this model: according to the comparison of accuracy level table, the accuracy level of this prediction model is level 3, and the level is low. As a prediction model, it is obviously unqualified.

The second-order weakening of the original data makes the data present a certain regularity to establish a prediction model:

$$XD^2 = XDD = [x(1)d^2, x(2)d^2, \ldots\ldots x(n)d^2] \quad (3.3)$$

$$x(k)d^2 = \frac{1}{n-k+1}[x(k)d + x(k+1)d + \ldots\ldots + x(n)d], k = 1,2,\ldots\ldots,n \quad (3.4)$$

Substitute the original traffic data of the ship into:

$$x^{(0)} = [x^{(0)}(t)/432,397,358,326,298,312,281] \quad (3.5)$$

After the second order weakening, we get a set of different sequences:

IOP Conf. Series: Materials Science and Engineering **792** (2020) 012003 doi:10.1088/1757-899X/792/1/012003

$$X^{(0)}D^2, X^{(1)}D^2 = (295,268,254,228,216,231) \quad (3.6)$$

New sequence $X^{(1)}$ by accumulation

$$X^{(1)} = [X^{(1)}(1), X^{(1)}(2), X^{(1)}(3), X^{(1)}(4), X^{(1)}(5), X^{(1)}(6)] = (295,268,254,228,216,231) \quad (3.7)$$

Finally, the whitening equation is obtained:

$$\frac{dX^{(1)}}{dt} + aX^{(1)} = u, \quad (3.8)$$

Two parameters a and u are obtained:

$$Y_n = \begin{bmatrix} 268 \\ 254 \\ 228 \\ 216 \\ 231 \end{bmatrix}, \quad B = \begin{bmatrix} -477 & 1 \\ -784 & 1 \\ -1287 & 1 \\ -1465 & 1 \\ -1794 & 1 \end{bmatrix} \quad (3.9)$$

a=-0.013376, u=316.243201

Get the prediction model:

$$x^{-(1)}(t+1) = -21324.567884e^{-0.032543*k} + 21365.567884 \quad (3.10)$$

$$X^{(0)} = [X^{(0)}(1), X^{(0)}(2), X^{(0)}(3), \ldots\ldots X^{(0)}(6)] = (295, 268, 254, 228, 216, 231) \quad (3.11)$$

3.3. Establish absolute correlation

Calculation $X^{(0)}$ and $X^{(1)}$ absolute correlation, That is to establish the absolute correlation between the original data and the data processed by the second-order weakening operator.

Set the original data as $X^{(0)}$, The data processed by the second order weakening operator is $X^{(1)}$, Establish the absolute correlation between the two data, $X^{(0)}$ and $X^{(1)}$ the absolute correlation coefficient between them is K, if $k_0 > 0$, then $k > k_0$ the model is very reasonable and effective.

If $k = 0.914 \geq 0.9$ is calculated by the calculation method of absolute correlation degree, then the model can be checked to be qualified. After the model is determined to be qualified, the formula can be obtained

$$x^{-(1)}(t+1) = -23303.763654e^{-0.012654*k} + 23773.763654 \quad (3.24)$$

This formula can be used as the prediction model of grey system. Through this model, we can calculate the prediction of ship traffic accidents in Qingdao sea area and Liaoning sea area in 2018, 2019 and 2012 respectively (1), and they are 265, 245 and 231 respectively.

3.4. Model prediction test

The second-order weakening operator is used to process the original data of ship traffic accidents in 2012-2018, and then the prediction model is used to obtain the accident prediction data of 2019, 2020 and 2021 respectively, with the prediction value data. See table 5 for the prediction value dataStyle and spacing

Table 5. Prediction results of GM (1,1) model.

Particular year	2012	2013	2014	2015	2016	2017	2018	2019	2020	2021
K		1	2	3	4	5	6	7	8	9
$X^{(1)}$	245	561	1042	1212	1463	1644	1832			
$X^{(2)}$		241	233	225	217	220	218			
Original value	319	308	296	286	283	288	284			
Predicted value (1)							296	291	286	
Predicted value (2)								278	265	254

4. Conclusion s

By combining the data of Liaoning sea area and Qingdao sea area with the prediction analysis calculated by the grey system model, it is a very simple and effective method to build GM (1,1) model to predict maritime traffic accidents. However, only two sea area traffic accident data are selected for superposition, the number of samples is small, and this model generally has high prediction accuracy under short-term and constant conditions. Therefore, in order to predict the sea area with frequent accidents accurately and stably in a long term, this model needs to be further optimized and improved.

References

[1] DENG J L.Control problems of grey systems[J].Systems & ControlLetters,1982,1(5):288-294.

[2] Tao Jun, Li Yaodong. Research on ship traffic supervision under poor visibility in Zhanjiang port [J]. Zhujiang water transport, 2017,27 (15): 72-73.

[3] Tu Haiyang, Yuan Zhangxin. Research on simulation system of ship traffic accident recurrence [J]. Shanghai Institute of ship transportation science, 2001.

[4] DEHDARA F,YAPB S F,NAGHAVIC M S,et al. Charting thefuture global status of oil and natural gas using grey forecasting[J]. Institutions and Economies,2016,8(3):105-125

[5] Kong Xue, Wang Li, Feng Yihua. Application status and Prospect of Grey Prediction GM (1,1) model [J]. Journal of Qilu University of technology, 2018,32 (6): 49-53.

[6] XU Q Y,YANG M L,LIU W D,et al.Prouduct form feature evolution forecasting based on IGMBPM madel[J].Computer-Aided Design and Applications,2016,13(4):431-439

[7] Deng Julong. Basic method of grey system [M]. Wuhan: Huazhong University of Technology Press, 1987

[8] Liu Sifeng, Xie Naiming. Grey system theory and its application [M]. Beijing: Science Press, 2018.

IOP Conf. Series: Materials Science and Engineering **792** (2020) 012004 doi:10.1088/1757-899X/792/1/012004

Risk hierarchy analysis of international railway container transportation

Ling Gao[1]*, Kaixin Chen[2]

[1] College of Economics and Management, Fuzhou University, Fuzhou, 350106, China

[2] College of Economics and Management, Fuzhou University, Fuzhou, 350106, China

* Corresponding Author's e-mail: gr003043@163.com

Abstract: The implementation of the "the Belt and Road" strategy has deepened the ex change of goods between countries and regions along the "Silk road economic belt". Until April 2019, China launched more than 15,000 trains of the international container trains. Because there are many kinds of transportation goods, many transportation segment, the whole transport process has to face multiple risk threats. According to the "5M" theory, this paper constructs a transportation risk indicator system with 5 first-level indicators and 28 second-level indicators, applicate AHP to calculate the weights of all indicators, selected 17 indicators. Using ISM, the risk hierarchy structure is determined and a three-level interpretative structure model is established. The first level is the surface factor, which including three indicators, if the surface risk factors are identified, it means that the entire transportation system is facing a serious risk crisis. The second level is the middle level factor, which including 11 indicators. The middle factors are risk factors for releasing the crisis signal which is usually closely related to the macro policy changes and the organizational capabilities and resilience of each link. The third level is the deep-seated factor, including three indicators, which requires risk managers to take countermeasures of risk warning and prevention.

1. Introduction

With the expansion of the "the Belt and Road" initiative, the exchanges between countries and regions along "the Belt and Road" have been deepened. Since the opening of the international railway container transportation lines in 2011 to April 2019, more than 60 cities in China have opened international railway express trains, of which the number has exceeded 15,000. The transported cargo includes mechanical equipment, complete automobiles and spare parts, daily necessities, household appliances, tires, electronic products, luggage and stationery, agricultural products, integrated circuits and so on.

The international railway transportation spans the Asian-European continent, and there are many countries with many transportation links. The transportation process faces multiple risk threats. The risk factors from different channels have different impacts and influences on the whole process of transportation. However, there are complex causal logical relationships between these risk factors. In other words, the changes that occur during the transportation of cargo are a complex system process that is a combination of multiple risk variables. Therefore, under the premise of clarifying the basic relationship between various risk factors, this paper uses the explanatory structure model to analyze the multi-level hierarchical structure of international container freight transport risks, which is highly intuitive and can understand the risks at a glance through the hierarchical chart.

2. Literature review

Ma, X.L. et al. based on risk source analysis, establish an accident probability model based on the ordered Logit model and a diamond-based accident consequence evaluation model to obtain the comprehensive risk level of each section of the transportation route [1]. Wang, Z. et al. established a dynamic risk assessment model for dangerous cargo transportation on urban expressway, focusing on the impact of traffic flow parameters on the probability of dangerous cargo accidents and the quantity of population exposure [2]. Wu, J.Z. et al. analyzed the risk factors of road transportation of dangerous cargo, and uses fuzzy comprehensive evaluation method to establish a risk evaluation model to reflect the weak points of safety management [3]. Su, J.X. constructed a four-layer explanatory structure model to study the influencing factors of the inland river dry bulk shipping freight rate and the relationship between various factors [4]. Sadeh, Garkaz. analyzed the impact of various factors of medical services and hospitality services on perceived value, satisfaction and loyalty according to ISM [5] Guo, D.et al. selected nine risk factors, constructed an explanatory structure model and analyzed the hierarchical relationship between the risk factors of fresh cold chain logistics [6]. Lei, K. et al. constructed a multi-level hierarchical analytical structure model to explore the structure and hierarchical relationship of risk factors in multimodal transport networks [7]. As can be seen from the above literature, using the ISM model to analyze the risk of container transportation is a relatively new attempt.

3. Screening risk evaluation indicators

4M theory pointed out that the occurrence of accidents is mainly related to four factors: human, environment, management and Facility equipment. Once an accident happens, it is a multi-faceted combination utility of the above four factors. However, the cargo on the international railway transportations transport has a wide variety, different structure and the required transport conditions. Therefore, the occurrence of transportation accidents is also highly correlated with the cargo factors. When considering the factors of transportation accidents, not only the four factors of human, environment, management, and Facility equipment should be considered, but also the nature of the cargo should be considered. Therefore, the cause of the accident is extended from "4M theory" to "5M theory".

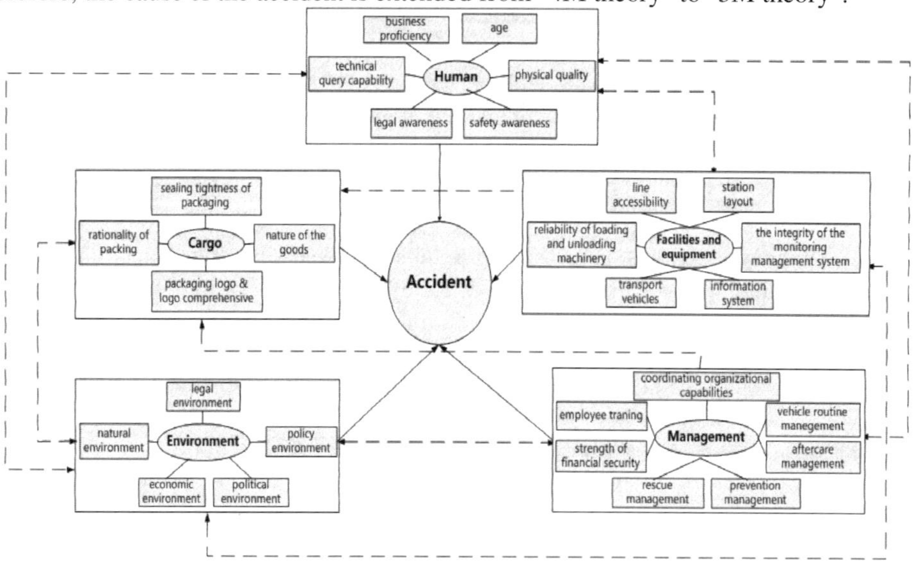

Figure 1. Causality diagram of international railway transportation accidents.

Fig.1 can be shown that:

- Changes in any of the five factors can cause an accident.
- The five factors are interrelated and unified. For example, an inappropriate behavior of a person may cause the vehicle to be in an unsafe state, increasing the probability of an accident;

external environmental factors may affect the safety status of the cargo, and may increase the probability of an accident and slow the speed of rescue.

Based on *Fig.1*, an international railway transportation risk evaluation indicators system with 5 first-level indicators and 28 second-level indicators can be constructed. We call this the "initial indicator system". Although this indicator system comprehensively covers the factors affecting the risk of cargo transportation, too many indicators are not only easy to repeat, but also weaken the importance of important indicators, so it must be deleted.

We select AHP to conduct this research and carry out a questionnaire survey. The questionnaire was distributed to universities and research institutes, as well as middle and high-level personnel engaged in international freight transportation .The measures of questionnaire distribution mainly includes mailing questionnaires, field research ,and intercept interviews .The scale of the importance of the indicator is judged to be 5 levels, that is, "very important", "important", "general", "unimportant" and "very unimportant", and their corresponding values are 5, 4, 3, 2, 1 respectively.

We distributed 46 questionnaires and collected 34 copies, of which 32 were valid questionnaires. Using spss 22.0, performed the Cronbach reliability analysis. The Cronbach coefficient value was 0.829, which was highly reliable.

Calculating questionnaire scores, making a risk evaluation score table, and constructing each element comparison matrix. This paper calculates the weight of each indicator and the consistency ratio of each element matrix by using the AHP. The calculation results are shown in the table1).

Table 1. First-level indicators weight and consistency ratio.

	First-level Indicators	Weights	Consistency ratios
Risk evaluation indicators system	A_1 (Human factors)	0.3785	0.056
	A_2 (Facility equipment factors)	0.1177	0.019
	A_3 (Environmental factors)	0.2364	0.041
	A_4 (Management factors)	0.2345	0.03
	A_5 (Cargo factors)	0.0737	0.0077

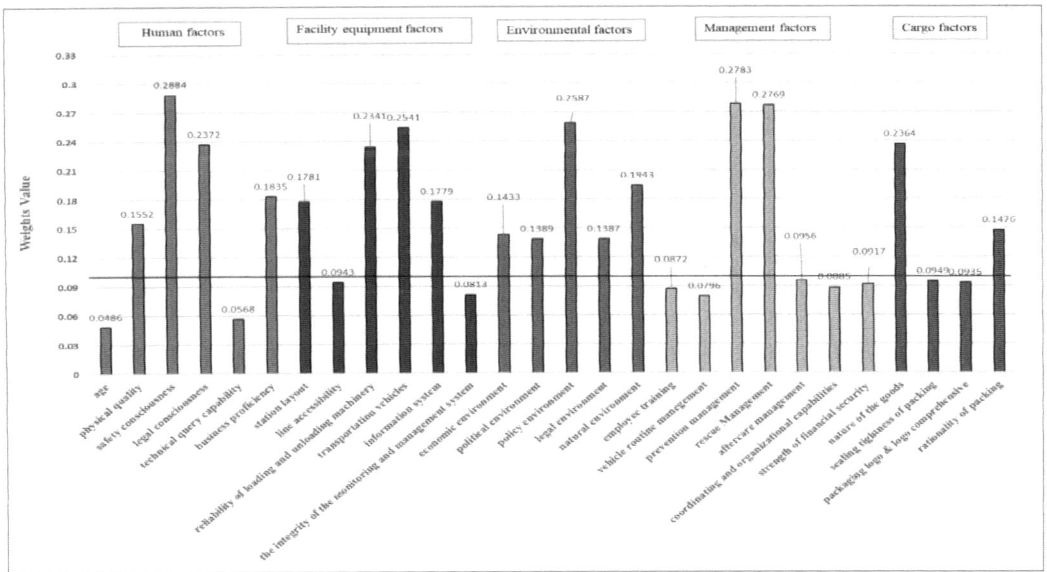

Figure 2. Second-level indicators weight.

By calculation , all indicators $CR \leq 0.1$, accord with the consistency test. Because the number of indicators below a certain criterion layer more than nine indexes, the critical point of the index weight can be taken as 0.1. Therefore, according to the index weight value of *Fig.2*, deleting the indicators with weights less than 0.1 and retaining the indicators with weights greater than 0.1 and screening out 17 indicators that avoid the risk concentration and index repeatability.

Table 2. Risk indicator screening result.

Deleted risk indicators	age, technical query capability, line accessibility, the integrity of the monitoring and management system, employee training, vehicle routine manegement, coordinating and organizational capabilities, strength of financial support, sealing tightness of packaging, rehabilition disposal capacity, packaging logo and logo comprehensive
Retained risk indicators	physical quality, safety awareness,law awareness,business proficiency, station layout, reliability of loading and unloading machinery, transportation tools, information systems, economic environment, political environment, policy environment, legal environment, natural environment, prevention management, rescue Management, the nature of the goods,rationality of packaging

4. Application of Interpretative Structural Model

4.1. Establish a binary relationship of risk indicators

Eight experts were invited to compare 17 risk indicators in pairs and analyze whether each indicator has a direct impact on other indicators. Usually V、A、O are used to indicate the relationship between the factors, V means that S_i affects S_j; A means that S_j affects S_i; and O means S_i and S_j do not affect each other.

Table 3. Binary relationship of risk indicators in international railway transportation.

	S_{17}	S_{16}	S_{15}	S_{14}	S_{13}	S_{12}	S_{11}	S_{10}	S_9	S_8	S_7	S_6	S_5	S_4	S_3	S_2	S_1
S_1 physical quality										V							
S_2 safety awareness				V													
S_3 legal awareness				V													
S_4 business proficiency										V							
S_5 reliability of loading and unloading machinery				V													
S_6 station layout							A		A								
S_7 information system				V													
S_8 transportation vehicles				V													
S_9 economic environment																	
S_{10} political environment							V										
S_{11} policy environment																	
S_{12} legal environment				V													
S_{13} natural environment			V														
S_{14} prevention management	A																
S_{15} rescue management		A															
S_{16} nature of the goods																	
S_{17} packing of the goods																	
T cargo transportation risk																	

4.2. Establish adjacency matrix

The set of elements is defined as $S=\{S_i \mid i=1,2,\cdots n\}$, and the matrix of the basic binary relationship between the elements in the set S is the adjacency matrix $C=(c_{ij})_{m\times n}$.

According to formula (1),

$$C = \left\{c_{ij} \left| S_i \vec{R} S_j = 1 \Rightarrow c_{ij} = 1, S_i \vec{R} S_j = 0, \Rightarrow c_{ij} = 0, i, j = 1, 2, ..., n; i \neq j \right.\right\} \quad (1)$$

"1" stands for relationship, and "0" stands for no relationship, establishing adjacency matrix C for risk factors of international railway cargo transportation.

C=

	S_1	S_2	S_3	S_4	S_5	S_6	S_7	S_8	S_9	S_{10}	S_{11}	S_{12}	S_{13}	S_{14}	S_{15}	S_{16}	S_{17}
S_1	0	0	0	0	0	0	0	1	0	0	0	0	0	0	0	0	0
S_2	0	0	0	0	0	0	0	0	0	0	0	0	0	1	0	0	0
S_3	0	0	0	0	0	0	0	0	0	0	0	0	0	1	0	0	0
S_4	0	0	0	0	0	0	0	1	0	0	0	0	0	0	0	0	0
S_5	0	0	0	0	0	0	0	0	0	0	0	0	0	1	0	0	0
S_6	0	0	0	0	0	0	0	0	0	0	0	0	0	0	0	0	0
S_7	0	0	0	0	0	0	0	0	0	0	0	0	0	1	0	0	0
S_8	0	0	0	0	0	0	0	0	0	0	0	0	0	1	0	0	0
S_9	0	0	0	0	0	1	0	0	0	0	0	0	0	0	0	0	0
S_{10}	0	0	0	0	0	0	0	0	0	0	1	0	0	0	0	0	0
S_{11}	0	0	0	0	0	1	0	0	0	0	0	0	0	0	0	0	0
S_{12}	0	0	0	0	0	0	0	0	0	0	0	0	0	1	0	0	0
S_{13}	0	0	0	0	0	0	0	0	0	0	0	0	0	0	1	0	0
S_{14}	0	0	0	0	0	0	0	0	0	0	0	0	0	0	0	0	0
S_{15}	0	0	0	0	0	0	0	0	0	0	0	0	0	0	0	0	0
S_{16}	0	0	0	0	0	0	0	0	0	0	0	0	0	0	1	0	0
S_{17}	0	0	0	0	0	0	0	0	0	0	0	0	0	1	0	0	0

4.3. Solve reachable matrix

The reachable matrix is used to describe the extent to which a certain path can be reached between different nodes in a directed connection graph.

The reachable matrix Q is calculated according to the formula (2).

$$Q = (C+I)^{i+1} = (C+I)^{i} \neq (C+I)^{i-1} \neq \ldots \neq (C+I), i = 1, 2, \ldots, n. \quad (2)$$

I is a C-order unit matrix.

Q=

	S_1	S_2	S_3	S_4	S_5	S_6	S_7	S_8	S_9	S_{10}	S_{11}	S_{12}	S_{13}	S_{14}	S_{15}	S_{16}	S_{17}
S_1	1	0	0	0	0	0	0	1	0	0	0	0	0	1	0	0	0
S_2	1	1	0	0	0	0	0	0	0	0	0	0	0	1	0	0	0
S_3	0	0	1	0	0	0	0	0	0	0	0	0	0	1	0	0	0
S_4	0	0	0	1	0	0	0	1	0	0	0	0	0	1	0	0	0
S_5	0	0	0	0	1	0	0	0	0	0	0	0	0	1	0	0	0
S_6	0	0	0	0	0	1	0	0	0	0	0	0	0	0	0	0	0
S_7	0	0	0	0	0	0	1	0	0	0	0	0	0	1	0	0	0
S_8	0	0	0	0	0	0	0	1	0	0	0	0	0	1	0	0	0
S_9	0	0	0	0	0	1	0	0	1	0	0	0	0	0	0	0	0
S_{10}	0	0	0	0	0	1	0	0	0	1	1	0	0	0	0	0	0
S_{11}	0	0	0	0	0	1	0	0	0	0	1	0	0	0	0	0	0
S_{12}	0	0	0	0	0	0	0	0	0	0	0	1	0	1	0	0	0
S_{13}	0	0	0	0	0	0	0	0	0	0	0	0	1	0	1	0	0
S_{14}	0	0	0	0	0	0	0	0	0	0	0	0	0	1	0	0	0
S_{15}	0	0	0	0	0	0	0	0	0	0	0	0	0	0	1	0	0
S_{16}	0	0	0	0	0	0	0	0	0	0	0	0	0	0	1	1	0
S_{17}	0	0	0	0	0	0	0	0	0	0	0	0	0	1	0	0	1

4.4. Hierarchical division of influence factors

According to formulas (3), (4), (5), the reachable set R, the premise set H, and the common set G required for structural interpretation are filtered in the reachable matrix Q, and then the hierarchical relationship of elements are decomposed according to the following process.

$$R = R(S_i) = \left\{R(S_i) = \left\{S_i \left| q_{ij} = 1, j = 1, 2, ..., n\right.\right\} \left| i = 1, 2, ..., n\right.\right\} \quad (3)$$

$$H = H(S_i) = \left\{H(S_j) = \left\{S_i \left| q_{ij} = 1, i = 1, 2, ..., n\right.\right\} \left| j = 1, 2, ..., n\right.\right\} \quad (4)$$

$$G = G(S_i) = \left\{G(S_i) = R(S_i) \cap H(S_j) \left| i = j = 1, 2, ..., n\right.\right\} \quad (5)$$

• Take the intersection of $R(S_i)$ and $H(S_i)$ to get the highest level element set L_1, which is the termination set:

$$L_i = \{S_k \mid R(S_K) \cap H(S_K) = R(S_K), K = 1, 2, \ldots n\}, i = 1, 2, \ldots, n \tag{6}$$

- The elements in the set L_1 are eliminated in $R(S_i)$ and $H(S_i)$, and the set L_2 is continuously searched in the remaining sets.
- This loops until you get the set L_n of the lowest level.

In this order, a collection of elements at each level is obtained:

$$L_1 = \{S_6, S_{14}, S_{15}\}$$
$$L_2 = \{S_2, S_3, S_5, S_7, S_8, S_9, S_{11}, S_{12}, S_{13}, S_{16}, S_{17}\}$$
$$L_3 = \{S_1, S_4, S_{10}\}$$

4.5. Determine the hierarchy diagram

According to the obtained various factor levels, the reachable matrix Q is reordered to obtain a standard form hierarchical structure matrix Q'.

$$Q'=\begin{bmatrix}
 & S_6 & S_{14} & S_{15} & S_2 & S_3 & S_5 & S_7 & S_8 & S_9 & S_{11} & S_{12} & S_{13} & S_{16} & S_{17} & S_1 & S_4 & S_{10} \\
S_6 & 1 & 0 & 0 & 0 & 0 & 0 & 0 & 0 & 0 & 0 & 0 & 0 & 0 & 0 & 0 & 0 & 0 \\
S_{14} & 0 & 1 & 0 & 0 & 0 & 0 & 0 & 0 & 0 & 0 & 0 & 0 & 0 & 0 & 0 & 0 & 0 \\
S_{15} & 0 & 0 & 1 & 0 & 0 & 0 & 0 & 0 & 0 & 0 & 0 & 0 & 0 & 0 & 0 & 0 & 0 \\
S_2 & 0 & 1 & 0 & 1 & 0 & 0 & 0 & 0 & 0 & 0 & 0 & 0 & 0 & 0 & 0 & 0 & 0 \\
S_3 & 0 & 1 & 0 & 0 & 1 & 0 & 0 & 0 & 0 & 0 & 0 & 0 & 0 & 0 & 0 & 0 & 0 \\
S_5 & 0 & 1 & 0 & 0 & 0 & 1 & 0 & 0 & 0 & 0 & 0 & 0 & 0 & 0 & 0 & 0 & 0 \\
S_7 & 0 & 1 & 0 & 0 & 0 & 0 & 1 & 0 & 0 & 0 & 0 & 0 & 0 & 0 & 0 & 0 & 0 \\
S_8 & 0 & 1 & 0 & 0 & 0 & 0 & 0 & 1 & 0 & 0 & 0 & 0 & 0 & 0 & 0 & 0 & 0 \\
S_9 & 1 & 0 & 0 & 0 & 0 & 0 & 0 & 0 & 1 & 0 & 0 & 0 & 0 & 0 & 0 & 0 & 0 \\
S_{11} & 1 & 0 & 0 & 0 & 0 & 0 & 0 & 0 & 0 & 1 & 0 & 0 & 0 & 0 & 0 & 0 & 0 \\
S_{12} & 0 & 1 & 0 & 0 & 0 & 0 & 0 & 0 & 0 & 0 & 1 & 0 & 0 & 0 & 0 & 0 & 0 \\
S_{13} & 0 & 0 & 1 & 0 & 0 & 0 & 0 & 0 & 0 & 0 & 0 & 1 & 0 & 0 & 0 & 0 & 0 \\
S_{16} & 0 & 0 & 1 & 0 & 0 & 0 & 0 & 0 & 0 & 0 & 0 & 0 & 1 & 0 & 0 & 0 & 0 \\
S_{17} & 0 & 1 & 0 & 0 & 0 & 0 & 0 & 0 & 0 & 0 & 0 & 0 & 0 & 1 & 0 & 0 & 0 \\
S_1 & 0 & 1 & 0 & 0 & 0 & 0 & 0 & 1 & 0 & 0 & 0 & 0 & 0 & 0 & 1 & 0 & 0 \\
S_4 & 0 & 1 & 0 & 0 & 0 & 0 & 0 & 1 & 0 & 0 & 0 & 0 & 0 & 0 & 0 & 1 & 0 \\
S_{10} & 1 & 0 & 0 & 0 & 0 & 0 & 0 & 0 & 0 & 1 & 0 & 0 & 0 & 0 & 0 & 0 & 1
\end{bmatrix}$$

Remove the transfer relationship between elements in the standard form hierarchy matrix Q' and its relationship to obtain the level matrix Q''.

$$Q''=\begin{bmatrix}
 & S_6 & S_{14} & S_{15} & S_2 & S_3 & S_5 & S_7 & S_8 & S_9 & S_{11} & S_{12} & S_{13} & S_{16} & S_{17} & S_1 & S_4 & S_{10} \\
S_6 & 0 & 0 & 0 & 0 & 0 & 0 & 0 & 0 & 0 & 0 & 0 & 0 & 0 & 0 & 0 & 0 & 0 \\
S_{14} & 0 & 0 & 0 & 0 & 0 & 0 & 0 & 0 & 0 & 0 & 0 & 0 & 0 & 0 & 0 & 0 & 0 \\
S_{15} & 0 & 0 & 0 & 0 & 0 & 0 & 0 & 0 & 0 & 0 & 0 & 0 & 0 & 0 & 0 & 0 & 0 \\
S_2 & 0 & 1 & 0 & 0 & 0 & 0 & 0 & 0 & 0 & 0 & 0 & 0 & 0 & 0 & 0 & 0 & 0 \\
S_3 & 0 & 1 & 0 & 0 & 0 & 0 & 0 & 0 & 0 & 0 & 0 & 0 & 0 & 0 & 0 & 0 & 0 \\
S_5 & 0 & 1 & 0 & 0 & 0 & 0 & 0 & 0 & 0 & 0 & 0 & 0 & 0 & 0 & 0 & 0 & 0 \\
S_7 & 0 & 1 & 0 & 0 & 0 & 0 & 0 & 0 & 0 & 0 & 0 & 0 & 0 & 0 & 0 & 0 & 0 \\
S_8 & 0 & 1 & 0 & 0 & 0 & 0 & 0 & 0 & 0 & 0 & 0 & 0 & 0 & 0 & 0 & 0 & 0 \\
S_9 & 1 & 0 & 0 & 0 & 0 & 0 & 0 & 0 & 0 & 0 & 0 & 0 & 0 & 0 & 0 & 0 & 0 \\
S_{11} & 1 & 0 & 0 & 0 & 0 & 0 & 0 & 0 & 0 & 0 & 0 & 0 & 0 & 0 & 0 & 0 & 0 \\
S_{12} & 0 & 1 & 0 & 0 & 0 & 0 & 0 & 0 & 0 & 0 & 0 & 0 & 0 & 0 & 0 & 0 & 0 \\
S_{13} & 0 & 0 & 1 & 0 & 0 & 0 & 0 & 0 & 0 & 0 & 0 & 0 & 0 & 0 & 0 & 0 & 0 \\
S_{16} & 0 & 0 & 1 & 0 & 0 & 0 & 0 & 0 & 0 & 0 & 0 & 0 & 0 & 0 & 0 & 0 & 0 \\
S_{17} & 0 & 1 & 0 & 0 & 0 & 0 & 0 & 0 & 0 & 0 & 0 & 0 & 0 & 0 & 0 & 0 & 0 \\
S_1 & 0 & 0 & 0 & 0 & 0 & 0 & 0 & 1 & 0 & 0 & 0 & 0 & 0 & 0 & 0 & 0 & 0 \\
S_4 & 0 & 0 & 0 & 0 & 0 & 0 & 0 & 1 & 0 & 0 & 0 & 0 & 0 & 0 & 0 & 0 & 0 \\
S_{10} & 0 & 0 & 0 & 0 & 0 & 0 & 0 & 0 & 0 & 1 & 0 & 0 & 0 & 0 & 0 & 0 & 0
\end{bmatrix}$$

4.6. Construction Interpretative Structural Model

The logical structure relationship between the various factors is clarified, and the explanatory structure model of each risk indicator is obtained, as shown in the *Fig.3*.

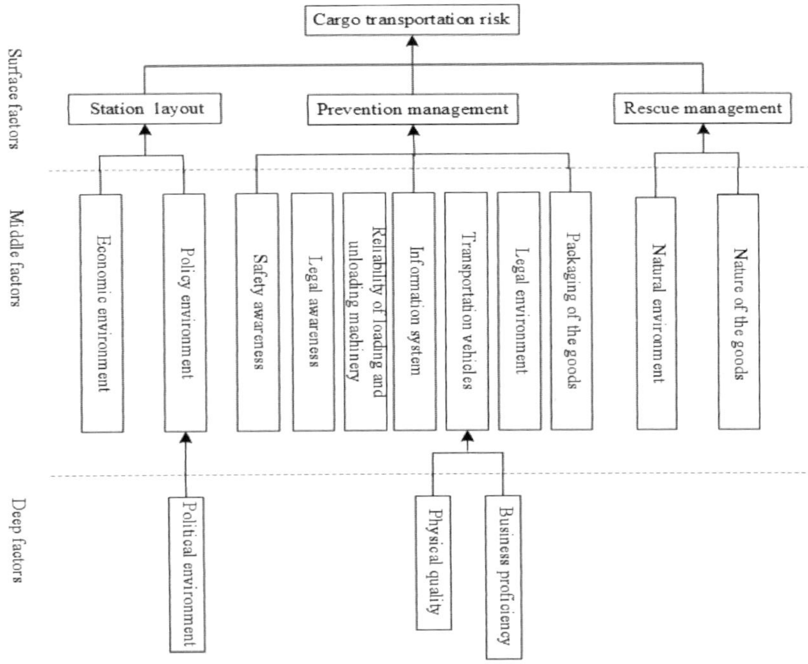

Figure 3. Hierarchical structure diagram of the risk

5. Conclusions

Fig.3 shows that the hierarchical structure model of the correlation coupling of the factors of the international railway cargo transportation risk. Using ISM to analyze the hierarchical relationship of cargo transportation risk factors can help to clarify the complex relationship between risk factors.

The first level is the surface factor, including the station layout, prevention management and rescue management, directly acting on the operating system. If the surface risk factors are identified, it means that the entire transportation system is facing a serious risk crisis, and corrective measures must be taken immediately.

The second level is the middle factors, including the economic environment, policy environment, safety awareness, legal awareness, reliability of loading and unloading machinery, information systems, transportation vehicles, legal environment, packaging of goods, natural environment and the nature of goods, that indirectly act on the operating system. The middle factors are risk factors for releasing the crisis signal which is usually closely related to the macro policy changes and the organizational capabilities and resilience of each link. That is the key and difficult point to control the risk of the overall transportation system.

The third level is a deep factor, including political environment, physical quality and business proficiency. Risk managers need to continuously strengthen themselves and take countermeasures in the aspect of risk warning and prevention. The deep factors transmit energy through the middle layer to the surface layer under the stimulation of uncertainty and induce the risk crisis of the international railway transportation at different levels, it brings a lot of negative energy to the organizational operation. Therefore, international railway cargo transportation related enterprises need to do well in risk prevention and control measures at all levels to ensure the safe and efficient transportation of goods.

Acknowledgements

This research was financially supported by the National Social Science Foundation of China (NSSFC). (Grant No. 14 BJY 136).

References

[1] Ma, X.L., Liu, Y.J., Lu, J. (2018) Risk assessment of dangerous cargo transportation on urban roads, Chinese Journal of Safety Science (05): pp.179-184.
[2] Wang, Z., Tao, J., Wang, W.J. (2018) Dynamic risk assessment of dangerous cargo transportation on urban expressway, Journal of Chongqing Jiaotong University (Natural Science) (07): pp.6.
[3] Wu, J.Z., Fan, W.J. (2015) Research on risk assessment system for dangerous cargo road transport, Highway transportation technology (12): pp.6-11.
[4] Su, J.X. (2019) Research on influencing factors of Inland river dry bulk water transportation price based on explanatory structural model, Logistics Technology (05); pp.6-9.
[5] Sadeh, Garkaz. (2019) Interpretive structural modeling of quality factors in both medical and hospitality ser vices in the medical tourism industry, Journal of Travel & Tourism Marketing, Taylor journal (02):pp.253-267.
[6] Guo, D., Pan, C.H., Lee, H.Y., Fu, M. (2018) Logistics risk factors of intelligent fresh cold chain based on ISM, Chinese Business Theory (32): pp.16-17.
[7] Lei, K., Zhu, X.N., Hou, J.F. (2015) Structural analysis of risk factors in multimodal transportation network based on ISM model, Journal of Safety & Environment (06): pp.162-165.

The Analysis and Model of AC Switch Indication Circuit

Zhang Lu

Beijing National Railway Research & Design Institute of Signal & Communication Group Co., Ltd. Beijing, 100070, China

zhanglu@crscd.com.cn

Abstract. Alternating current (AC) switch machine is usually a few kilometers away from the signal building. When the cable X1 of AC switch indication circuit breaks, it indicates that the relay does not release. Based on this problem, AC switch indication circuit is analysed, a method for solving the distributed capacitance of cable is given, and a simulation model of the AC switch indication circuit is established. The test was carried out at Hami south station, and the validity of the simulation model was verified.

1. Introduction

The switch indication circuit is a circuit that reflects the switch position to the signal building[1]. The switch indication circuit is used for not only supervision, but also interlocking, so it must take more perfect fail-safe measure.

However, in an interlocking test conducted by Chengdu Railway Bureau, it was found that the switch indication relay was not released when the cable X1 disconnected with the outdoor cable box, and the cable length was 4.429km. Coincidentally, 16 groups of alternating current (AC) switch in Urumqi Railway Bureau, whose control cable length was more than 2km, was found the same problem. It does not conform to the fail-safe principle[2].

In this paper, the simulation model of AC switch indication circuit is established to analyze the situation of cable X1 broken in this circuit, and verify the engineering improvement schemes.

2. AC switch indication

In the AC switch indication circuit, the primary side voltage of the transformer is AC 220V, and the secondary side voltage is AC 110V. DBJ is the switch normal indication relay, and FBJ is the switch reverse indication relay. Therefore, the positive half cycle circuit and the negative half cycle circuit of the power supply are analyzed respectively when analyzing the working principle of the switch normal indication. The switch reverse indication is the same.

The positive half cycle circuit of the power supply is shown in red line of Figure 1[3]. The current flow direction is as follows: transformer +→R1→1DQJ23-21→2DQJ131-132→1DQJF13-11→2DQJ111-112→switch machine contact 31-32→switch machine contact 15-16→rectifier diode Z→resistance R2→switch machine contact 33-34→motor coil U→motor coil W→1DQJ11-13→transformer -.

The negative half cycle circuit of the power supply is shown in blue line of Figure 1. The current flow direction is as follows: transformer -→R1→1DQJ23-21→2DQJ131-132→DBJ4-1→switch machine contact 11-12→motor coil V→motor coil W→1DQJ11-13→transformer +.

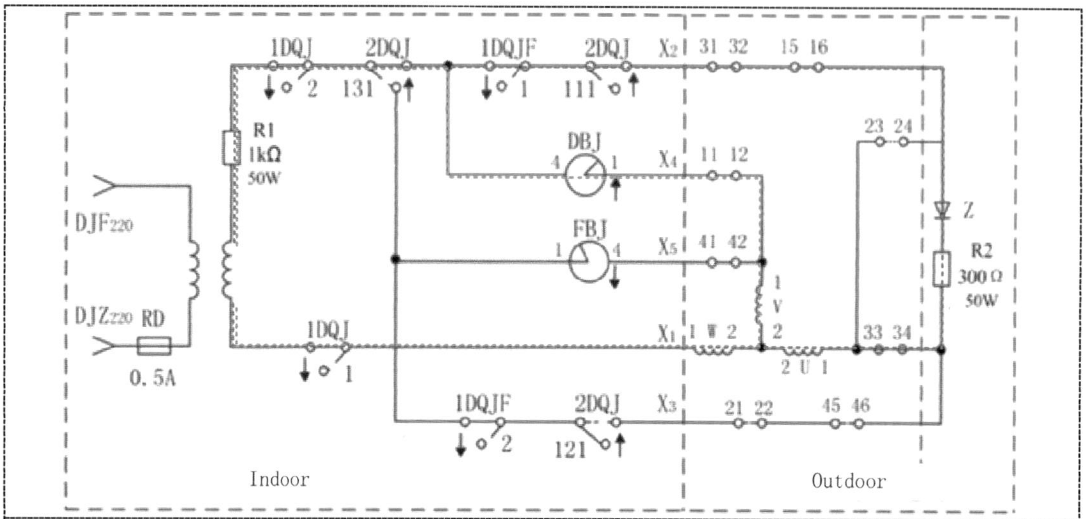

Figure 1. AC switch indication circuit

In order to analyze the AC switch indication circuit conveniently, Separate the AC switch normal indication circuit and switch reverse indication circuit, leave out the contact of relay and switch machine, and retain electrical components in the circuit. The equivalent circuit is shown in Figure 2.

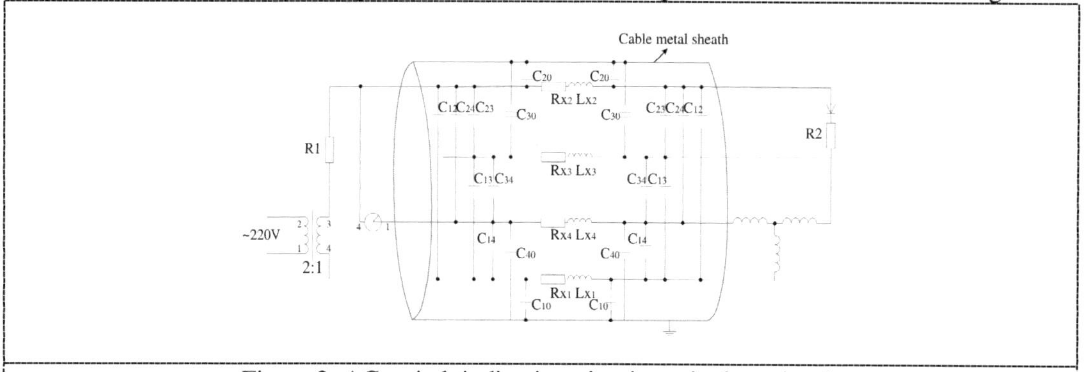

Figure 2. AC switch indication circuit equivalent circuit

It can be seen from Figure 2 that there are distributed capacitances between cable core wire and core wire, between core wire and metal sheath[4]. When the circuit X1 is disconnected with the circuit, the AC current will detour to the terminal relay through the distributed capacitance. The longer the cable length is or the more cable the parallel core wires are, the larger the distributed capacitance is and then the greater the current flow to the relay is. If the distributed capacitance is large enough, the relay would keep the energized state when the cable X1 disconnected with the circuit.

3. Circuit analysis and model

The AC switch indication circuit include three parts, namely, signal source, transmission channel and terminal impedance.

3.1. Signal source

The signal source part is composed of indoor AC signal source, transformer and outdoor rectifier circuit. The 50 Hz AC signal source loaded on the terminal relay after diode rectification is a periodic signal with high positive half cycle peak voltage and low negative half cycle peak voltage. The measured waveform is shown in Figure 3.

Discrete Fourier transform is applied to the voltage signal, and the spectrum characteristics is shown in Figure 4. The signal energy distribution is mainly concentrated in the 0 ~ 300 Hz wide-band signal, in which the 50Hz AC component accounts for a large proportion. In the process of transmission, this signal makes the influence of cable distribution capacitance intensified, and also causes the impedance at both ends of the relay to increase.

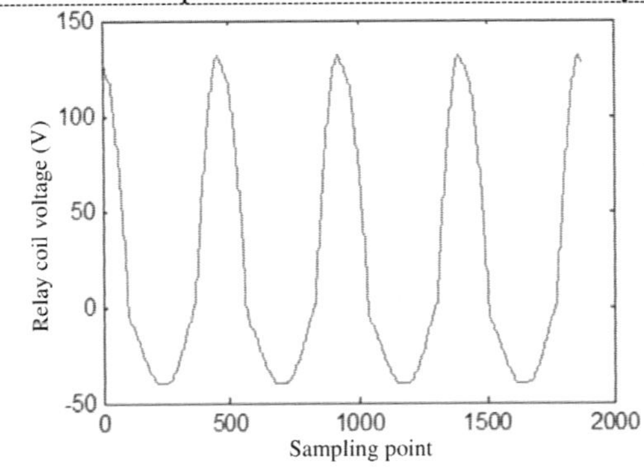

Figure 3. Measured waveform of relay coil voltage signal

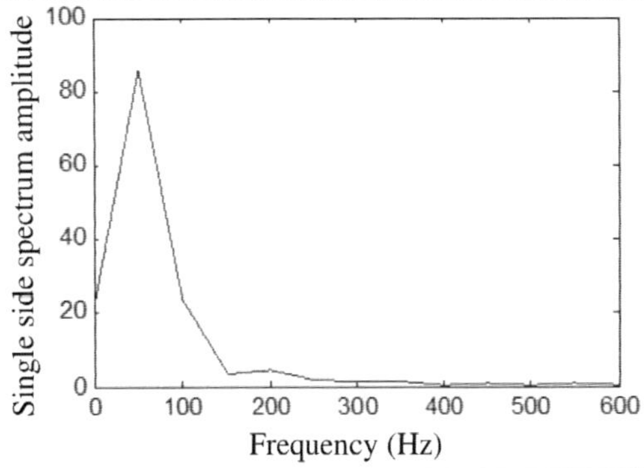

Figure 4. Frequency spectrum analysis diagram of relay coil voltage signal

3.2. Transmission channel

The transmission channel is composed of outdoor signal transmission cable and switch machine motor coil. The parameters of the switch machine can be obtained directly from the test. The resistance and inductance parameters of cable also can be tested by conventional method. The cable distributed capacitance will be analyzed in detail, because the railway digital signal cable has the characteristics of many specifications, asymmetric or irregular arrangement of core wires.

In any circuit, a distributed capacitance is formed between two insulated conductors with differential pressure. Therefore, there is a distributed capacitance between the core wires of the cable and between the core wire and the metal sheath of the cable. The distributed capacitance value depends on the geometry of the cable, the length of the cable and the insulating material.

3.2.1. Test and solution method of the distributed capacitance between core wire and metal sheath

One core wire in the cable is selected as the tested object A, and all the other core wires are short circuited as test object B (short circuited to eliminate the distributed capacitance between other core wires), and then measuring capacitance C_A C_B C_{AB} with capacitance tester, as shown in Figure 5

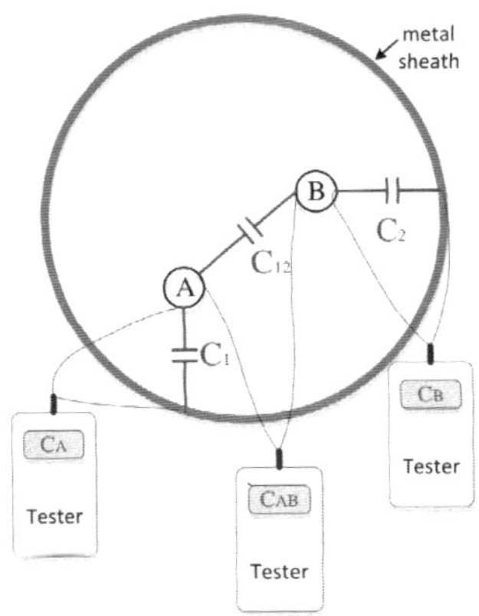

Figure 5. Test method for the distributed capacitance between core wire and metal sheath

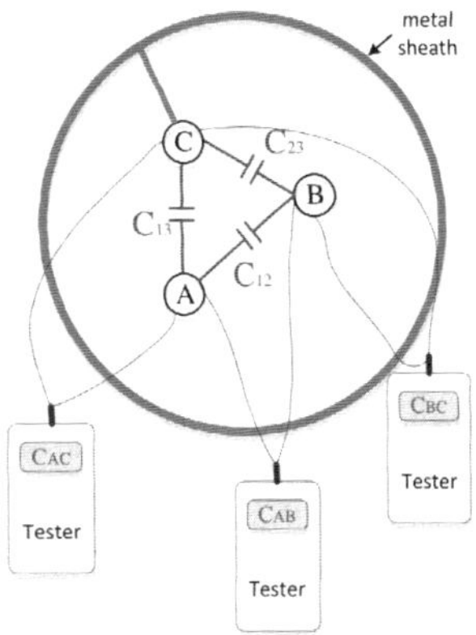

Figure 6. Test method for the distributed capacitance between core wire and core wire

IOP Conf. Series: Materials Science and Engineering **792** (2020) 012005 doi:10.1088/1757-899X/792/1/012005

According to equations (1), the value of C_1 can be calculated. C_1 is the distributed capacitance between the tested object core wire A and the metal sheath..

$$\begin{cases} C_A = C_1 + \dfrac{C_{12} \times C_2}{C_{12} + C_2} \\ C_B = C_2 + \dfrac{C_{12} \times C_1}{C_{12} + C_1} \\ C_{AB} = C_{12} + \dfrac{C_2 \times C_1}{C_2 + C_1} \end{cases} \quad (1)$$

3.2.2. Test and solution method of the distributed capacitance between core wire and core wire

Select one core wire in the cable as tested object A and another core wire in the cable as tested object B. All the other core wires are short circuited as tested object C (short circuited to eliminate partial capacitance between other core wires), and tested object C should be short circuited with metal sheath (short circuited to eliminate capacitance between test object C and metal sheath), and then measuring capacitance C_{AB} $C_{AC}C_{BC}$with capacitance tester, as shown in Figure 6

According to equations (2), the value of C_{12} can be calculated. C_{12} is the distributed capacitance between the tested object core wire A and the tested object core wire B.

$$\begin{cases} C_{AB} = C_{12} + \dfrac{C_{13} \times C_{23}}{C_{13} + C_{23}} \\ C_{AC} = C_{13} + \dfrac{C_{12} \times C_{23}}{C_{12} + C_{23}} \\ C_{BC} = C_{23} + \dfrac{C_{13} \times C_{12}}{C_{13} + C_{12}} \end{cases} \quad (2)$$

According to the above distributed capacitance test and solution method, 42 core wires railway digital signal cable is selected for test, and the layout of core wires are shown in Figure 7[5]. The solution results of distributed capacitance between core wire and metal sheath are shown in Table 1, and the distributed capacitance between core wire and core wire is shown in Table 2.

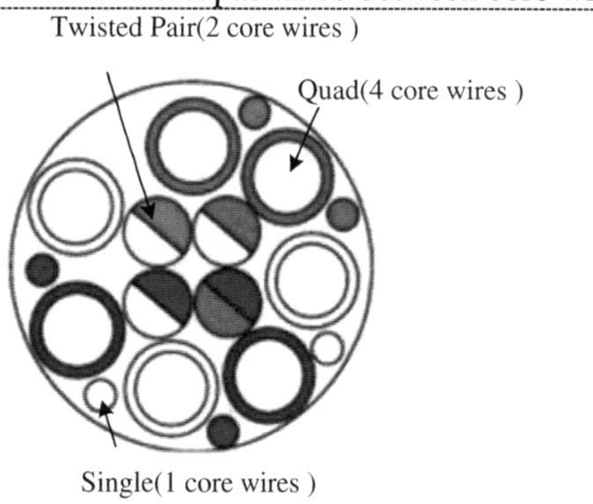

Figure 7. 42 core wires cable section diagram

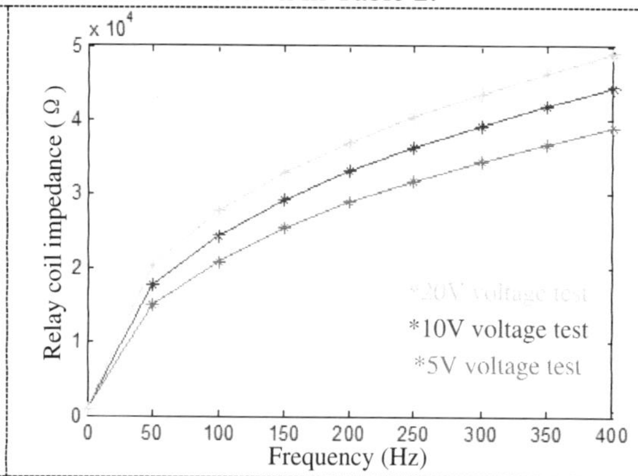

Figure 8. Relay coil impedance versus signal voltage and frequency

IOP Conf. Series: Materials Science and Engineering **792** (2020) 012005 doi:10.1088/1757-899X/792/1/012005

Table 1. The distributed capacitance between core wire and metal sheath

	Distributed capacitance（nF/km）
Between a core wire of quad and metal sheath	6~7
Between a core wire of Twisted pair and metal sheath	0~0.2
Between single core wire and metal sheath	17~18.5

Table 2. The distributed capacitance between core wire and core wire

	Distributed capacitance（nF/km）
Between two adjacent core wires in one quad	17~18
Between two diagonal core wires in one quad	3~4
Between two core wires in two adjacent Quad	1.2
Between two core wires in two nonadjacent Quad	0
Between a core wire in a quad and red single	6
Between red single and green single	0.15
Between two core wires in one twisted pair	20

3.3. terminal impedance

The relay used in turnout indication circuit is JPXC-1000 bias relay, which is composed of two coils with direct current (DC) resistance of 500 Ω in series. The coil shows great impedance under the aforementioned wide-band signal source, and the impedance of the coil changes with the AC signal voltage and frequency, as shown in Figure 8.

Therefore, when setting the simulation parameters, the impedance of the relay coil is optimized to make the parameters of the simulation model more close to the actual working impedance of the relay.

3.4. Simulation model establishment

In Matlab / Simulink, the simulation analysis model of AC switch representation circuit is established, as shown in Figure 9. Among them, the signal source parts are the red 220 V power supply, BD1-7 transformer and diode module; the transmission channel parts are the green transmission cable module and switch machine coil UVW module; the terminal impedance is the blue JPXC-1000 relay module.

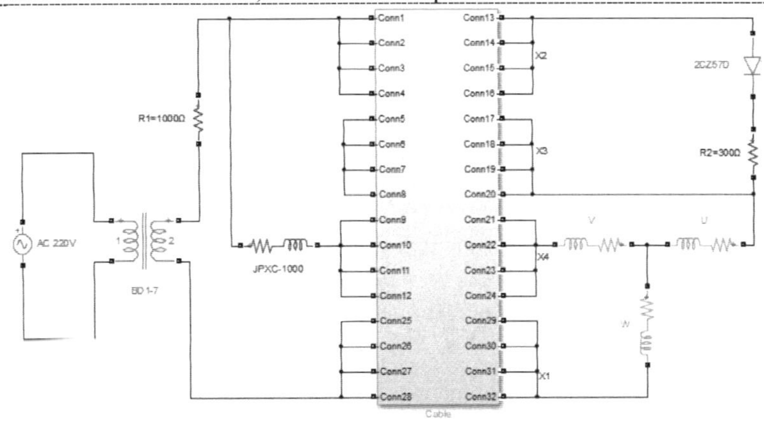

Figure 9. Simulation analysis model of AC switch indication circuit

4. Experimental verification

In order to verify the validity of the simulation model, field test was carried out at Hami South Station of Urumqi Railway Bureau.

The test object is the 104# switch (cable length 5.21km) in Hami south station. The voltage of the relay coil under the normal condition and the condition of cable disconnected with circuit situation were tested under the condition of four core wires in parallel, three core wires in parallel, two core wires in parallel and single core wire.

Under normal and disconnected test records, the relay coil voltage root mean square (RMS) is shown in Table 4. Under normal condition, the errors between test results and simulation calculation results are not more than 1%; under X1 disconnected with circuit condition, the errors between test results and simulation calculation results are not more than 7%, which verifies the validity of simulation model.

Table 3. Comparison between simulation results and test results

Cable paralleling	Relay coil voltage RMS (V)					
	Normal condition			X1 disconnected with circuit		
	test	simulation	error	test	simulation	error
four core wires in parallel	21.3	21.4	<1%	13.6	13.6	0
three core wires in parallel	21	21.1	<1%	10.7	11.2	4.6%
two core wires in parallel	20.5	20.5	0	9.04	9.2	1.8%
single core wire	19.1	19.1	0	5.62	6.0	6.8%

5. Conclusion

(1) Through the analysis of the AC switch indication circuit, it is determined that the root cause of X1 disconnection relay not released is the cable distribution capacitances.

(2) This paper presents a method of measuring and solving the distributed capacitance of railway signal cable, which can obtain the distributed capacitance between any two cores or between any core wire and metal sheath in any number of cables (including cables with complex, asymmetric and irregular core wire arrangement).

(3) The simulation analysis model of the AC turnout representation circuit is established and compared with the field measurement data to verify the validity of the model. It can guide the simulation verification of engineering improvement scheme

References

[1] He WQ. (2011) Switch control circuit. In: Wei JY. 6502 Electric Central Circuit. China Railway Publishing House, Beijing. pp. 96-110.

[2] Wang Y, Zhang L, Yang L.(2018)Research on the influence of cable distributed capacitance on the AC switch indication circuit. Railway Signalling & Communication Engineering, 15(1):8-11.

[3] Lin YY, Fu XM, Hong G. (2012) Switch control circuit. In: Cui ZW, Xu Q, Wei JY. 6502 Electrical centralized circuit Atlas. In: China Railway Publishing House, Beijing. pp. 36-39.

[4] Qiu GY. (2000) Capacitance element. In: Liu JY. Circuit. Higher Education Press, Beijing. pp. 9-12.

[5] TB/T 3100.1-2017 Railway digital signalling cable-Part 1: General specifications.

IOP Conf. Series: Materials Science and Engineering **792** (2020) 012006 doi:10.1088/1757-899X/792/1/012006

Establishment of criteria for determining preventive maintenance requirements of cement pavement

Chen Qiqi*, Wang Guanhu, Long Xiaoyong, Lei Jichao

School of Aeronautical Engineering, Air Force Engineering University, Xi'an 710038, China

*Corresponding author's e-mail: kgdcqq@163.com

Abstract: In order to obtain the criteria for preventive maintenance of cement pavement in military airport, based on the basic assumption of flatness-dynamic load interaction in the decay process of cement pavement, derivative inflection point is adopted. The method solves the relationship between the International roughness index (*IRI*) and the pavement damage index (*L*), and obtains the *IRI* critical value of the cement pavement when the *L* decay is the fastest. After the analysis and verification, the judgment standard of the cement pavement *IRI* is obtained. This paper analyzes the related research results of the preventive maintenance standard range of pavement condition index (*PCI*), summarizes and concludes the judgment standard of pavement *PCI*, and gets the *PCI-L* relation based on *L-IRI* and *PCI-IRI* relation, and finally gets the judgment standard of *L*. Based on the current situation that our army still mainly adopts the three-meter ruler for flatness test and a large number of measured data, a three-meter ruler average gap and a pavement *IRI* correlation relationship are established, which realizes the conversion between the two. Based on the current situation of less accumulation of performance test data of cement pavement in military airport, the damage condition and flatness data of several cement pavement in similar environment are fitted and analyzed by using the method of space generation time, and the specific parameters of the model are obtained.

1. Introduction

Preventive maintenance is an advanced maintenance concept, which means to reasonably and take corresponding maintenance measures for the existing pavement system on the premise that the pavement is in good condition. However, the maintenance work of the airport pavement of our army is still in the stage of "either bad or repair, neither bad nor repair". The time of overhaul or reconstruction usually comes ahead of schedule. Therefore, it is of great practical significance to carry out the research on the judgment standard of preventive maintenance of cement pavement in military airport.

At home and abroad, the research on preventive maintenance standards is mainly focused on asphalt pavement, and some research results have been obtained[1]. Wang Xiangfeng proposed the road condition standard for preventive maintenance of ordinary highway asphalt pavement by analyzing the road damage status index and road driving quality index test data of typical road sections in Sichuan Province over the years[2]. Ziari, H adopts packet data processing method and establishes neural network model to obtain preventive maintenance standard of flexible pavement[3].

However, as the cement concrete pavement is a rigid pavement, its bending strength is low and the compressive strength is high, so its continuity and integrity are not as good as asphalt pavement.It is impossible to copy the research results of the preventive maintenance judgment standard of asphalt pavement completely. Therefore, it is necessary to carry out the research on the corresponding preventive maintenance judgment standard in combination with the characteristics of cement concrete pavement. So far, there has been relatively little research in this area. This paper intends to systematically study the criteria for preventive maintenance by using methods such as summary induction, derivative inflection point and regression analysis, and establish a criterion for quantitatively guiding the preventive maintenance of cement pavement in military airports.

2. Determination of preventive maintenance demand criteria

2.1. Cement pavement international roughness index judgment standard

2.1.1.Solution of IRI criterion based on derivative inflection point. Unevenness is the inherent characteristic of cement concrete pavement. When an aircraft travels on the unevenness pavement, it will generate additional dynamic load on the pavement. The more uneven the surface is, the higher the dynamic load is, the faster the damage rate of the surface will be. There is a critical value in the process of flatness decay, and the critical value is not the time when the damage degree is the most serious, but the time when the acceleration of the development of the damage is the most. If the critical value is obtained and corresponding preventive maintenance measures are taken in the vicinity, the attenuation of the pavement performance can be most effectively mitigated.

According to the research results of pavement performance at home and abroad, combined with the analysis of the actual changes of pavement performance in some areas of China, sun Lijun[4] proposed the classic pavement decay equation as follows

$$PPI = PPI_0\left\{1\text{-exp}\left[-(\alpha/y)^{\beta}\right]\right\} \qquad (1)$$

Where: *PPI* is the service performance index; PPI_0 is the initial service performance index; y is the road age; α and β are model parameters, the size of parameter α reflects the length of the road service life, and the value of parameter β determines the road decay mode. When α and β take appropriate values, the *PPI* decreases with the increase of y, and the rate of change shows a trend of first fast and then slow. That is, there exists a certain y value, so that *PPI* declines at the maximum rate. By using different (α, β) combinations, different road decay processes can be fitted.

Zhou wenxian[5] adopts this model to study the relationship between pavement *PCI* and pavement *IRI*. The relationship between the two is as follows

$$PCI = 100\left\{1\text{-exp}\left[-(7.54/IRI)^{2.82}\right]\right\} \qquad (2)$$

In this paper, with reference to the practices of Sun Lijun and Zhou Wenxue, the relationship between the pavement *IRI-L* is established.

$$L=100\left\{1\text{-exp}\left[-(a/IRI)^{b}\right]\right\} \qquad (3)$$

Where: a and b are model parameters.

Based on the current situation that our military still uses the three-meter ruler for flatness testing, this paper intends to use *IRI* to describe the flatness of the pavement. Therefore, the three-meter ruler and *IRI* test vehicle are used for the flatness test of several cement concrete pavements of the same type in our army, and the data of several groups of *IRI* and *S* are obtained. By fitting the data of *S* and *IRI*, the conversion relationship between them is obtained as follows.

$$IRI = 0.9043 \times S - 0.0606 \qquad (4)$$

R^2=0.8164. Its conversion relationship diagram is shown in Figure 1.

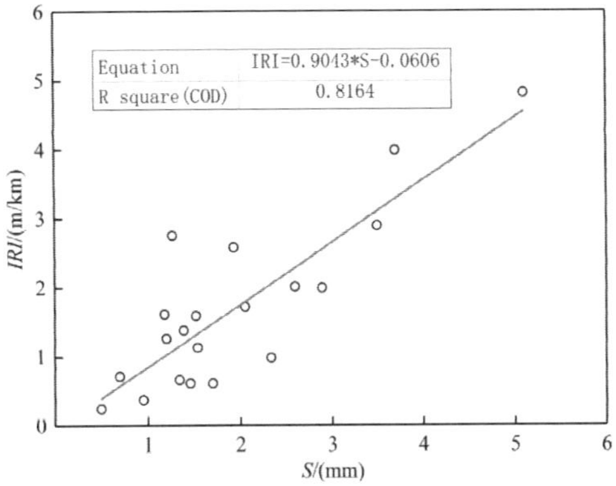

Figure 1. Conversion relationship between *IRI* and *S*

To establish an *L-IRI* relationship that conforms to the actual situation of our military airport pavement, it is necessary to obtain data on the long-term performance of the pavement. However, our military airport pavement has less historical data and the time series of data is not long. Therefore, this paper introduces the time-space conversion method, analyzes the performance data of several airport pavements with similar environment in our military, and obtains data of 50 sets of *L* and *S*. The formula (4) is used to convert *S* into *IRI*, and finally the fitting relationship between *L* and *IRI* is

$$L=100\left\{1-\exp\left[-(8.42/IRI)^{-2.46}\right]\right\} \quad (5)$$

R^2=0.8874. Its conversion relationship diagram is shown in Figure 2.

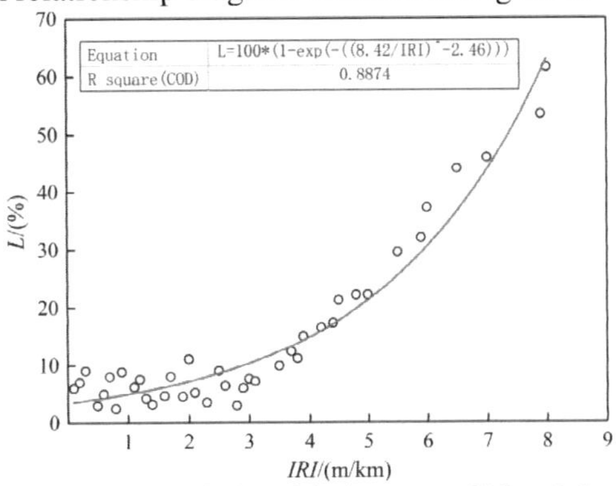

Figure 2. Relationship between *IRI* and *L*

In order to obtain the *IRI* corresponding to the point with the largest change rate of *L*, the second derivative of formula (5) is obtained, and the formula of the second derivative of *L* is

$$L'' = -100 * \alpha^{\beta} \exp\left(-\left(\alpha/IRI\right)^{\beta}\right) * \\ \left(\alpha^{b} \beta IRI^{-2\beta-2} - (b+1) IRI^{-\beta-2}\right) \quad (6)$$

Where L'' is the second derivative of the pavement damage index *L*, α =8.42, β =-2.46. To more clearly describe the *L-IRI* relationship, the relationship graph is shown in Figure 3.

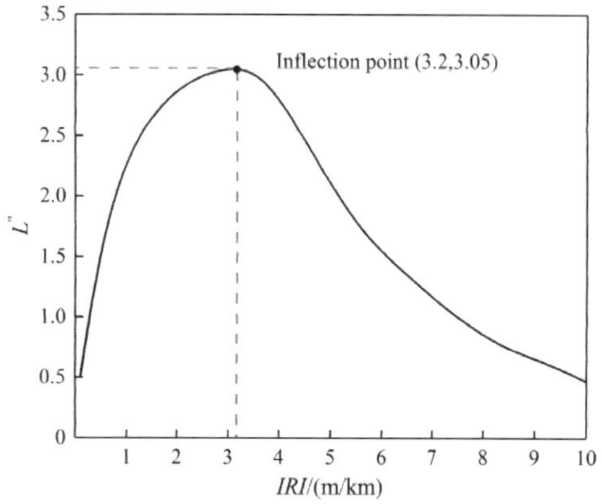

Figure 3. Relationship between the second derivative of *L* and *IRI*

As can be seen from Figure 3, the second derivative of the *L-IRI* function has an inflection point. When the *IRI* reaches this value of 3.2, the decay rate reaches a maximum, and the preventive maintenance effect is best.

2.1.2. Determination of IRI criteria for pavement. Flatness, as a macro index reflecting the flatness performance of pavement, usually its attenuation rate is relatively slow. If the upper limit value is selected to determine the preventive maintenance demand of pavement, the result may be contrary to the actual situation. For example, the flatness of a certain surface is relatively good, but other single indicators may have reached the threshold. If the prevention of preventive maintenance is determined by the upper limit of the flatness, it will lead to the conclusion that preventive maintenance is not required. Therefore, this paper suggests that only the lower limit standard should be set for the *IRI* criteria, which is used to make a preliminary judgment on the preventive maintenance needs of the pavement from the *IRI* perspective. Whether to prevent preventive maintenance in the end is also required to be combined with other indicators.

As can be seen from the previous section, the *IRI* at the fastest attenuation of the pavement is 3.2 m/km. Substituting it into equation (4), inversely calculating S=3.605mm, which roughly corresponds to the limit value of "good" and "medium" in the GJB 2264-1995 standard, which is consistent with the corresponding results of the highway. Therefore, this paper takes 3.2 m/km as the lower limit value of pavement *IRI* judgment standard, and finally the *IRI* preventive maintenance criterion is $IRI \leq 3.2$m/km.

2.2. *Judgment standard of pavement damage index*

2.2.1. Standard Range Analysis of Road Surface PCI. At present, there is no research on the criteria for preventive maintenance of pavement *PCI*, and the research on *PCI* preventive maintenance standards for asphalt pavement at home and abroad is relatively mature. Therefore, this paper summarizes the relevant research on the *PCI* preventive maintenance standard of road surface, refers to its experience method to determine the preventive maintenance standard, and puts forward the determination standard of *PCI* applicable to pavement. In order to combine with the actual situation of the army, the *PCI* of the pavement is transformed into the pavement damage index through a certain conversion relationship, and the judgment standard of the pavement damage index is established. Summarize the standard research results of some roads *PCI* at home and abroad, as shown in Table 1. The evaluation criteria for road damage are shown in Table 2.

Table 1. Standard range of road surface *PCI*

Researcher	*PCI*	Corresponding evaluation level
Wang Jiashu[1]	85-95	Excellent
Li Na[6]	60-85	Good, Medium
Zhang Liang[7]	70-90	Good, Excellent
He Guangbing[8]	82-88	Good, Excellent
Shanghai[9]	85-95	Excellent
Li Mingzheng[10]	70-95	Good, Excellent

Table 2. evaluation standard of road surface damage[11]

Evaluation level	Excellent	Good	Medium	poor	bad
PCI	≥85	70-85	55-70	40-55	≤40

According to the analysis of the research results of the above-mentioned pavement *PCI* standards, the general researchers think that when the pavement damage level is "excellent" and "good", it is the right time to carry out preventive maintenance. By averaging the upper and lower limits of the *PCI* preventive maintenance standards obtained by each investigator, the average standard range of *PCI* is (78.1, 90.3). As can be seen from Table 10, the upper limit value 90.3 is in the "excellent" range, and the lower limit value 78.1 is in the "good" range. The pavement is more stringent than the road surface requirements. Therefore, this paper strictly takes 75 as the lower limit and 95 as the upper limit, and finally determines that the range of the standard for the *PCI* of the pavement is (75,95).

2.2.2. *Conversion relationship between pavement PCI and L.* At present, there is no research on the relationship between the *PCI* and *L* transformation of the pavement. In order to establish the correlation between the *PCI* and the *L*, the *IRI* is introduced as an intermediate variable in this paper. The *PCI-L* relationship of the pavement is determined by the *IRI-L* relationship and the *IRI-PCI* relationship.

According to equation (5), the relationship of pavement *IRI-L* is established. Ling Jianming[12] carried out statistical regression analysis on the *PCI* value and *IRI* value of the survey units of the main runway of Shanghai Hongqiao Airport from 2000 to 2003. The correlation between *PCI* and *IRI* is obtained.

$$IRI = 23.007 - 4.684 \ln PCI \tag{7}$$

However, the formula is mainly applicable to asphalt pavement. At present, there is no literature to directly conduct a large number of regression statistics on the relationship between *PCI* and *IRI* on cement pavement. Therefore, based on the *PCI-IRI* relationship model of Ling Jianming, this paper takes 50 groups of *PCI-IRI* data from four military airports, corrects and fits the model, and obtains the *PCI-IRI* relationship formula of cement pavement.

$$IRI = 18.82 - 3.64 \ln PCI \tag{8}$$

R^2=0.9763. Its conversion relationship diagram is shown in Figure 4.

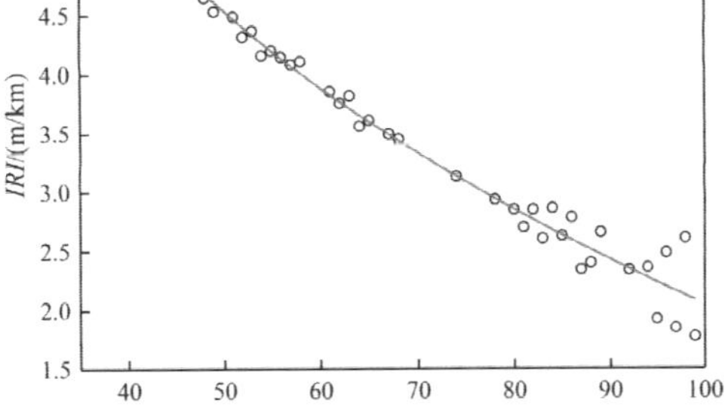

Figure 4. *PCI-IRI* Fitting correction diagram

Combining formula (5) (pavement *IRI-L* model) and formula (7) (pavement *PCI-IRI* model), through calculation, the relationship between *PCI-L* is obtained as follows

$$L = 121.9\text{-}26.11\ln PCI \tag{9}$$

R^2=0.9568. It shows that the fitting effect is good, and the transformation diagram is shown in Figure 5. Formula (9) can be used to convert the pavement *PCI* and *L*, which shows that it is feasible to deduce the *L* judgment standard according to the preventive maintenance standard of pavement *PCI*.

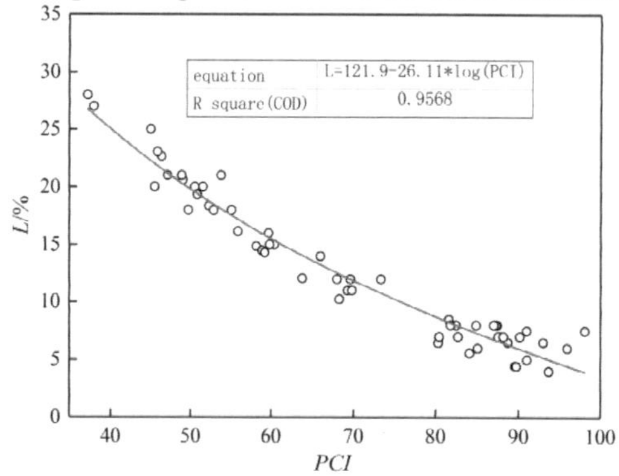

Figure 5. *PCI-L* Fitting correction diagram

2.2.3. Determination of judgment standard of pavement damage index. Through the above analysis, it is found that the standard range of pavement *PCI* is (75,95). Through the correlation transformation of pavement *PCI-L*, it is preliminarily found that the standard of pavement *PCI* is (3%, 10.5%). According to the classification standard of the appearance quality of cement concrete pavement of military airport (as shown in Table 3), the boundary value of "excellent" and "good" is 2.0%.

Table 3. criteria for damage index of military airport pavement

Evaluation level	Excellent	Good	Medium	poor	bad
Damage index (%)	<2	2 -5	5.1-15	15.1-20	>20

According to the *L-PCI* conversion relationship, the grading standards for the appearance quality of military airport pavements are relatively strict. If *L*=2.0% is used as the upper limit of the judgment standard, the preventive maintenance timing will be too advanced, which is not conducive to the determination of the optimal timing of preventive maintenance. Therefore, based on the actual situation, this paper takes 3.0% as the upper limit of the *L* criterion. Substituting the lower *IRI* standard value 3.4 obtained in section 2.1 into the formula(9), the corresponding *L*=10.16 %, which is consistent with the initially obtained 10.5 %, proves that the result is reasonable and reliable.This paper takes 10% as the lower limit of *L* criterion. The final determination standard of *L* is (3%, 10%).

3. Judgment standard of preventive maintenance demand

Based on the above analysis on the evaluation indexes of preventive maintenance from the International roughness index and pavement damage index, the judgment standard of preventive maintenance for cement pavement of military airport is determined, as shown in Table 4.

Table 4. Quantitative judgment standard for preventive maintenance demand of cement pavement

Preventive maintenance evaluation index		judgment standard
Flatness	International roughness index	≤3.2 m/km
Pavement damage	pavement damage index	(3%,10%)

4. Conclusion

Aiming at the current situation that the historical data of military airport pavement performance is less accumulated, the method of space generation time is used to fit and analyze the damage and flatness data of several cement pavements with similar environment in our military. The specific parameters of the model are obtained. On this basis, the research on the criteria for preventive maintenance of cement pavement in military airports is carried out.

Based on the basic assumption of the interaction between flatness and dynamic load in the course of the decay of cement concrete pavement, the derivative inflection point method is used to solve the *IRI-L* relation of pavement, and the critical value of the *IRI* of pavement with the fastest decay of *L* is obtained. Through the analysis and verification, the judgment standard of the *IRI* of pavement is obtained. This paper analyzes the research results related to the range of *PCI* preventive maintenance standards, obtains the judgment standard of pavement *PCI* through induction and analogy, and obtains the *PCI-L* relationship based on *L-IRI* and *PCI-IRI* relationship formula, and finally obtains the judgment standard of *L*.

Acknowledgments
Thank you very much for Professor Wang Guanhu's careful revision of this paper, which helped me point out many mistakes.

Reference

[1] Wang, J S. (2018) Research on Preventive Maintenance Effect of Asphalt Pavement Based on Data Statistics. Southeast University.
[2] Wang, X F. Yong, L M. (2017) Study on road condition standard and timing decision of preventive maintenance of highway asphalt pavement. Highway engineering, 42: 223-226.
[3] Ziari, H. Sobhani, J. Ayoubinejad, J. Hartmann, T. (2016) Prediction of *IRI* in short and long terms for flexible pavements: ANN and GMDH methods. International Journal of Pavement Engineering, 17: 776-788.
[4] Sun, L J. (2009) Road and Airport Facilities Management. China Communications Press, Beijing.
[5] Zhou, W X. (2007) At the most appropriate time to maintain the cement pavement. China Highway, 12: 84-85.
[6] Li, N. (2010) Research on predictive maintenance performance prediction model and decision model for highway asphalt concrete pavement. Beijing Jiaotong University.
[7] Zhang, L. (2011) Prediction of pavement condition index based on grey Markov Model. Science Technology and Engineering, 22: 5462-5465.
[8] He, G B. (2012) Research on preventive maintenance of highway asphalt pavement. Chang'an University.
[9] DG/TJ 08-2176 2015. (2015) Technical regulations for pre-maintenance of asphalt pavement. Shanghai Municipal Roads Bureau, Shanghai.
[10] Li, M Z. (2010) Selection matrix and programming of preventive maintenance measures for asphalt pavement. South China University of Technology.
[11] JTJ073.2-2001. (2001) Technical Specification for Maintenance of Highway Asphalt Pavement. Shanghai Highway Administration, Shanghai.
[12] Ling, J M. Liu, S F. Yuan, J. (2017) Application of IRI to evaluate the flatness of airport pavement. Journal of transportation engineering, 1: 20-27.

Research on the governance of parent-subsidiary company under the mixed ownership mode

YeZhu[1], WeiJiang[2,3*], Xinyu Shi[4] and Yuntian Dai[5]

[1]Strategic Department, Shanghai Aviation Electric Co., Ltd., Shanghai, 201100, China

[2]Flight College, Changzhou Institute of Technology, Changzhou, Jiangsu, 213032, China

[3]Jiangsu Nanfang Bearing Co., Ltd, Changzhou, Jiangsu,213163, China

[4]School of Mechanical Engineering, University of Bristol, Bristol, BS8 1QU, UK

[5]School of Mechanical Engineering, University of Birmingham, Birmingham B15 2TT, UK

[*]e-mail:jwei@czust.edu.cn

Abstract. As the core content of state-owned enterprises reform, Chinese state-owned enterprises have some problems in corporate governance for a long time, which restricts the process of the reform and development in these enterprises. As a breakthrough in deepening the reform of state-owned enterprises, the mixed-ownership reform has opened up a new theoretical research direction and reform practice space for state-owned enterprises to carry out and improve corporate governance. Under the framework of corporate governance and contemporary property rights theory and based on the analysis of the characteristics and existing problems of corporate governance in Chinese state-owned enterprise groups, this paper summarizes the framework and specific measures that can be taken by state-owned enterprises to improve corporate governance under mixed-ownership reform.

1. Introduction

1.1. Purpose of the Study

In September 2015, the CPC Central Committee and the State Council issued 'the guiding opinions on deepening the reform of state-owned enterprises', which points out the direction on how to improve the modern system, market management mechanism and state-owned assets management system.With the reform of state-owned enterprises entering a difficult period, the mixed-ownership reform (MOR) has become a significant breakthrough. The government aims to combine the advantages of state-owned enterprises and private enterprises through the MOR. Under the support of the national policy, some relative reform pilots are carrying out in a manner. However, after prompting the MOR, how to balance the interests of all shareholders and take advantages of asset and efficiency will be a crucial problem that enterprises will face when they change from a single shareholder to a diversified shareholder.

IOP Conf. Series: Materials Science and Engineering **792** (2020) 012007 doi:10.1088/1757-899X/792/1/012007

1.2. Research Content
The corporate governance in this paper is in a narrow sense. It is the institutional arrangement of functions and structures of the directorate, the rights of shareholders. This paper supposes that corporate governance should be established a set of static structure and dynamic mechanism about ownership and control rights for enterprises with independent corporate status, which mainly focuses on the following two aspects:

It is a problem with the ownership and distribution of control rights among shareholders. To be specific, it is also the power distribution and responsibility between controlling shareholders and small ones.

It is a problem with the ownership and distribution of the control right of the company between the shareholders and management, which is the power and responsibility relationship between the owners and the managers.

2. The general situation of corporate governance of state-owned enterprises in China
The corporate governance of state-owned enterprises in China has absorbed different characteristics of British, American, Japanese and German in the process of development, which forms a corporate governance model of state-owned enterprises with Chinese characteristics. As follows, there are three characteristics:

2.1. Establishing a modern enterprise system with independent legal personality
After many years of reform, the majority of the state-owned enterprises have formed self-employed legal persons with the nature of limited liability company and joint-stock limited company. At the same time, the management of managers in state-owned enterprises is also changing from administrative appointment to market-oriented employment.

2.2. Using the capital market to establish a diversified equity structure dominated by state-owned equity
In recent years, China has vigorously encouraged state-owned enterprises to raise funds by listing and other market-oriented ways and introduced private, personal and foreign capital into the equity structure, which significantly improves the operation efficiency and management level of the state-owned enterprises.

2.3. Coexistence of independent director system and board of supervisor system
In the process of the establishment of the board of directors, the state-owned enterprises in China use the experience of different corporate governance structures of the British, American, Japanese and German, which forms the coexistence of independent directors and the board of supervisors.

Although China has carried out much productive work in the field of state-owned enterprise reform, the governance level of state-owned enterprises has been significantly improved. However, the standard is restricted by historical reasons, market environment, industry characteristics, system background and other factors. There are still some problems in the corporate governance of state-owned enterprises, which are manifested explicitly in the following aspects:

2.3.1. Loss of owner
There are three links of principal-agent relationship between the ultimate owner and the operator in the state-owned enterprises in China: political principal-agent, administrative principal-agent and economic principal-agent. The long chain of principal-agent relationship not only increases the cost of the agency but also weakens the enthusiasm of the clients to supervise the agents. This is the problem of owner losing position.

2.3.2. Board of directors and managers with unclear rights and responsibilities
At present, the board of directors system of state-owned enterprises is not perfect. There are still some problems such as unclear responsibilities and distorted principal-agent relationship. In some state-owned enterprises, especially those directly controlled by the state-owned assets supervision department as the investor, the board of directors system only exists in the form, but never play a substantial role.

2.3.3. Lack of sufficient incentive and restraint mechanism
In terms of establishing an effective incentive and restraint mechanism, some Chinese state-owned enterprises have carried out much useful exploration, but there are still some obvious problems:

Firstly, the treatment level of managers in state-owned enterprises is embarrassed. On the one hand, the high salary of executives in some state-owned enterprises has caused the dissatisfaction of the public opinion; on the other hand, the average income level of managers in state-owned enterprises is still low, which has limited incentives to managers.Secondly, the salary structure of managers in state-owned enterprises is too single. Finally, it is the lack of effective restraint mechanism. Restricted by the underdeveloped manager labour market and the long-term concept of cadre management, managers in state-owned enterprises seldom feel the external pressure like managers in fully market-oriented enterprises.

3. Methodology
Corporate governance is a set of systematic organisational system design and implementation. Jensen (1993) divides the mechanism of corporate governance into decision-making mechanism, incentive mechanism and supervision mechanism [5].

In December 2015, the State-owned Assets Supervision and Administration Commission, the Ministry of Finance, and the Development and Reform Commission issued the Guiding Opinions on the Definition and Classification of the Functions of State-owned Enterprises, which defined state-owned enterprises as economic and public welfare. The primary purposes of commercial state-owned enterprises are to enhance the vitality of the state-owned economy, enlarge the functions of state-owned capital and realize the preservation and appreciation of state-owned assets. These enterprises implement commercial operations by market-oriented requirements. The main objectives of public welfare state-owned enterprises are to protect people's livelihood, serve the society, and provide public goods and services. The government can regulate the prices of necessary products or services.

According to the classification of state-owned enterprises in the guidance opinions and given the above discussion on the connotation of corporate governance and domestic and foreign governance experience, the following content summarizes the institutional arrangements and implementation rules of corporate governance from the three dimensions of decision-making, incentives and supervision mechanisms.

3.1. Decision Mechanism
The decision-making mechanism refers to the mechanism by which company power is distributed among and within corporate bodies such as shareholders' meeting, the board of directors, the board of supervisors, and the managerial level. It is the core of the incentive mechanism and the supervision mechanism. It is also the core of corporate governance.

The specific manifestations of the decision-making mechanism are as follows:

Firstly, it is reflected in the distribution of power among various organizations within the company. On the one hand, as a vital force of external governance, specific laws and rules have clearly defined the internal power levels of the company, but from the perspective of implementation, different objective conditions make the statutory power level and actual power implementation not always consistent. For example, when the ownership structure is highly decentralized, it will be difficult for shareholders to achieve absolute control over the management. Therefore, the establishment of an equity structure that meets corporate governance objectives and the external conditions that can support such an equity structure is the basis of corporate governance.

Secondly, it is reflected in the distribution of power within various institutions. Within the shareholders' meeting, these distributions can be the distributions of power between the controlling shareholder and minority shareholder; within the board of directors, they can be the distributions of power between independent directors and internal directors.

Thirdly, it is reflected in the decision-making procedure when exercising power. For example, shareholders' meetings are divided into different skills according to annual meetings, and temporary meetings; Voting methods are divided into direct voting, cumulative voting classified voting, etc. for different matters; there is the avoidance principle of connected shareholders and connected directors; there are right of ruling and veto which are set up for chairman.

3.2. Incentive and Restraint Mechanism

Incentive mechanism is a set of mechanisms and arrangements what the principals use to prompt the agents to take appropriate actions when the company's ownership and the rights of management are separated, to ensure the best interests of principals.

According to the specific form and function of the incentive, the incentive mechanism includes fixed salary, which is the primary labour return and income guarantee of the motivated person; performance bonuses related to short-term performance, which promote the motivated person to work hard to achieve the short-term income goal of the company; stock options or restricted stock related to long-term performance, which ensure that the behaviour of the motivated person can balance the short and long-term interests of shareholders; the retirement plan, which can cultivate the motivated person's long-standing loyalty to the company; promote the unification of ownership and control Management shareholding plan of the company; reputation incentives for professional managers, which can increase the value of the motivated person in the manager's market.

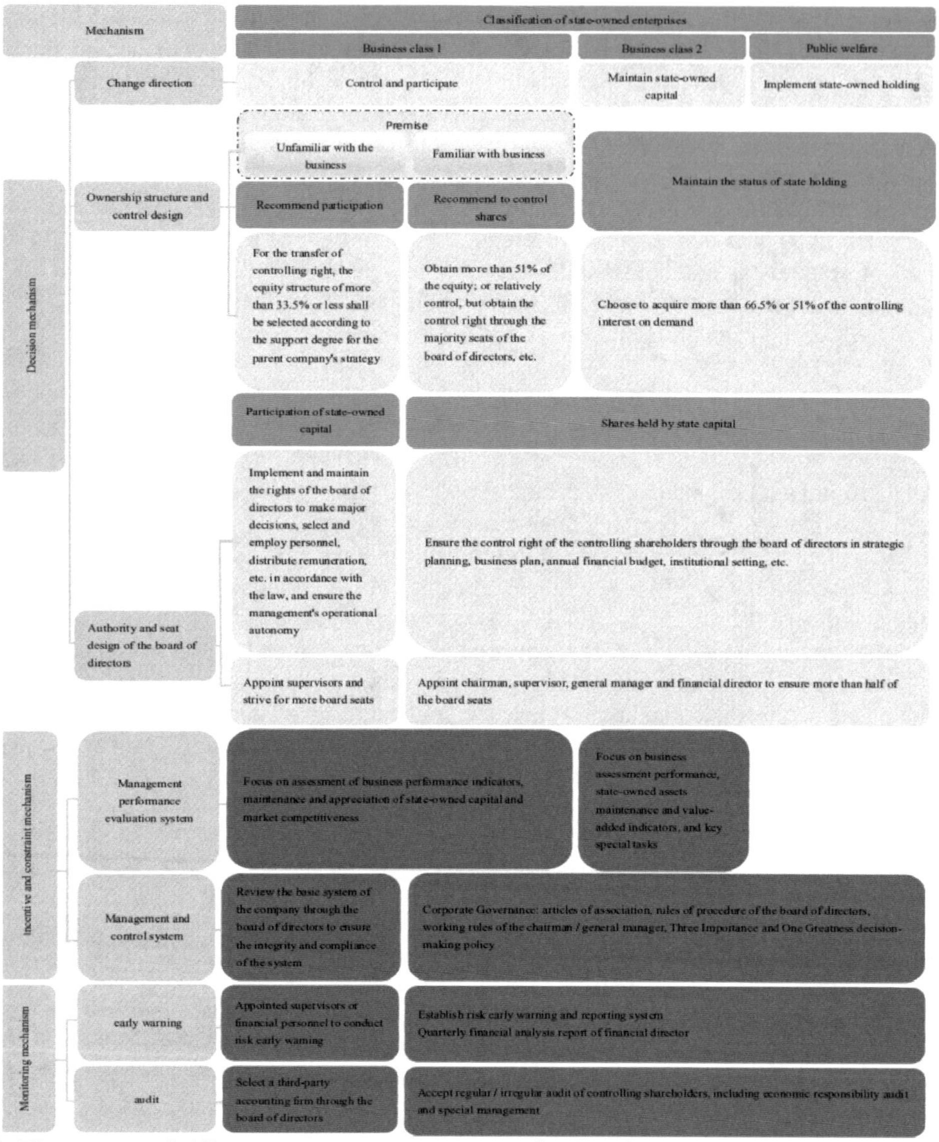

Figure 1. The parent-subsidiary company governance mechanism framework under a mixed *ownership model*

3.3. *Supervision Mechanism*

The supervision mechanism refers to the behaviours of stakeholders in reviewing, supervising and supervising the company's business operators in terms of business behaviour and business results. It is also the arrangement and action in the principal-agent relationship to ensure that the interests of principals are not harmed when the ownership and the rights of management are separated. Compared with the incentive mechanism, the supervision mechanism is more extensive and more rooted in corporate governance and is a direct reflection of the corporate governance mechanism.

The supervision mechanism of state-owned enterprises with mixed ownership can be designed from the perspective of internal and external aspects in combination with the setup of its governance structure. Internal supervision is mainly a supervision mechanism formed by shareholders, the board of directors, the board of supervisors and senior management. External supervision can be through a supervision mechanism formed by third-party auditing institutions and the public.

4. Conclusion

The supervision mechanism of state-owned enterprises with mixed ownership can be designed from the perspective of internal and external aspects in combination with the setup of its governance structure. Internal supervision is mainly a supervision mechanism formed by shareholders, the board of directors, the board of supervisors and senior management. This supervision mechanism includes shareholders' vertical supervision and supervision of the board of directors and senior managers. It also provides for the supervision of the board of directors and managers of the company by the board of supervisors and employee representatives. External supervision can be through a supervision mechanism formed by third-party auditing institutions and the public.

Acknowledgements
This work was financially supported by the Applied Basic Research Programs of Science and Technology Commission Foundation of Jiangsu Province (BK20150256) and the National Natural Science Foundation of China (51505126).

Reference

[1] Li, W. (2003) Corporate Governance of Commercial Banks: *Theoretical Models and Choices in China* [J]. Nankai Journal No.1.
[2] Dai, Y. (2012) *Research on Capital Structure Optimization of Listed Companies Based on Corporate Governance* [J]. Northeast Agricultural University.
[3] Zhang, W. (1999) *Enterprise Theory and Chinese Enterprise Reform* [M]. Beijing Peking University Press.
[4] Kevin, K, Steve, T, Mike, W. (2013) *Corporate Governance-Entrusted Responsibility, Corporate and International Comparison*. People's Posts and Telecommunications Press.
[5] Jensen, M. *The Modern Industrial Revolution, exit and the failure of internal control system*[J].Journal of Finance 1993,48:831-880.

IOP Conf. Series: Materials Science and Engineering **792** (2020) 012008 doi:10.1088/1757-899X/792/1/012008

Optimization of distribution path for community group buying considering customer satisfaction

Yong Wang[1], Wei Qiu[2*]

[1]Evergrande school of management, Wuhan university of science and technology, Wuhan, Hubei province, 430065, China

[2]Evergrande school of management, Wuhan university of science and technology, Wuhan, Hubei province, 430065, China

[*] Corresponding author's e-mail: 360662732@qq.com

Abstract. Aiming at the problems of low efficiency, high cost and low customer satisfaction in the logistics distribution of community group purchase enterprises. In this paper, a VRP model with minimum total cost including fixed cost, transportation cost and penalty cost is established under the constraint of satisfaction function of distribution time window. The designed genetic algorithm is used to solve the model. Through the example verification, the total cost function designed in this paper can effectively reduce the cost and obtain good satisfaction. The algorithm is feasible and effective, providing a reference for enterprise decision-making.

1. Introduction

For community group purchase logistics distribution, it is quite different from the traditional B2B distribution mode. At present, there are few literatures about community group purchase, and community group purchase enterprises usually adopt rough methods for distribution. Vehicle routing problem (VRPTW) with time window is one of them. Discussed the establishment of mathematical models for vehicle routing problems with time Windows [1]. Due to the np-hard characteristic, many scholars use heuristic algorithms to solve such problems. Include genetic algorithm [2], particle swarm optimization [3], and hybrid heuristic algorithm [4]. Many domestic scholars have done in-depth research on the vehicle routing problem of customer satisfaction with time window. Proposed a VRP optimization model for multi-model cold-chain logistics under the constraint of satisfaction [5]. Therefore, this paper first establishes the customer satisfaction function with service time as the variable. Then the total cost model including fixed vehicle cost, transportation cost and penalty cost is established.

2. Mathematical model of community group buying based on customer satisfaction

2.1. problem description

The specific problem can be described as that in a distribution center, all vehicles start from the distribution centre on the same day, know the location of the community leader point. Build the model based on the following assumptions:

- The location of the distribution center is fixed and distribution vehicles of the same type.
- The total demand of customers shall not exceed the stowage capacity limit of vehicles.
- The distribution vehicles run at a certain speed and the traffic is smooth.

- Each community should be served and only receive one service. The geographical location, demand and time range of receiving distribution service of each community leader are known.
- All delivery vehicles have time constraints when they arrive at the community.

2.2. symbol description

N is the set of community heads; K is the collection of distribution vehicles; Q is the maximum load of the vehicle; V is the average speed of the vehicle; d_{ij} is the distance between the previous community point i and the next community point j; p_c is the fixed cost of A car; p_r is the vehicle fuel cost per kilometer; ε_1 , ε_2 are the costs of vehicle waiting and tardiness per unit time. Decision-making scalar:
$x_{ijk} = \begin{cases} 1, & \text{Vehicle k goes from i to j} \\ 0, & \text{Otherwise} \end{cases}$ $y_{ik} = \begin{cases} 1, & \text{Customer i is delivered by vehicle k} \\ 0, & \text{Otherwise} \end{cases}$

2.3. satisfaction calculation

In this paper, the linear continuous time satisfaction function under the vehicle service time window is used to quantify customer time satisfaction CT_i, as shown in figure 1. CT_i is customer i satisfaction, and t_i is the moment when the vehicle arrives at customer i. The satisfaction function is expressed as

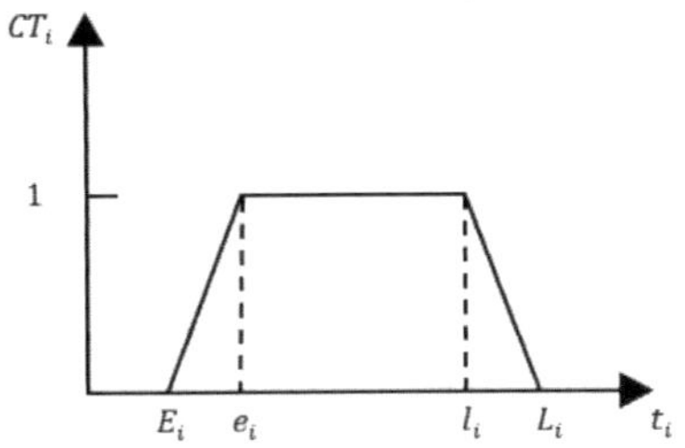

Figure 1. Time satisfaction

$$CT_i(t_i) = \begin{cases} 0, t_i \notin [E_i, L_i] \\ \dfrac{t_i - E_i}{e_i - E_i}, t_i \in [E_i, e_i) \\ \dfrac{L_i - t_i}{L_i - l_i}, t_i \in (l_i, L_i] \\ 1, t_i \in [e_i, l_i] \end{cases} \quad (1)$$

2.4. Mathematical model construction

Considering customer satisfaction, the distribution model is as follows:

$$Max \frac{1}{N} \sum_{i \in N} CT_i(t_i) \quad (2)$$

$$MinZ = \sum_{k=1}^{K} \sum_{j=1}^{N} x_{0jk} f_k + \sum_{k=1}^{K} \sum_{i=0}^{N} \sum_{j=0}^{N} d_{ij} x_{ijk} p_r + \varepsilon_1 \sum_{i=1}^{N} \max\{E_i - t_i, 0\} + \varepsilon_2 \sum_{i=1}^{N} \max\{t_i - L_i, 0\} \quad (3)$$

$$\sum_{i=1}^{N} q_i y_{ik} \le Q, k = 1, 2, ..., K \quad (4)$$

$$\sum_{k=1}^{K} y_{ik} = 1, i = 1, 2, \ldots, N \quad (5)$$

$$\sum_{j=1}^{N} x_{0jk} - \sum_{i=1}^{N} x_{i0k} = 0, \forall k \quad (6)$$

$$\sum_{i=0}^{N} x_{ijk} = y_{jk}, i = 1, 2, \ldots, N, \forall k \quad (7)$$

$$\sum_{j=0}^{N} x_{ijk} = y_{jk}, j = 1, 2, \ldots, N, \forall k \quad (8)$$

equation (2) is the objective function to maximize the average customer satisfaction; Equation (3) is the objective function that requires the minimum of fixed cost, transportation cost and penalty cost; Formula (4) limits the maximum carrying capacity of distribution vehicles; Formula (5) ensures that each colonel

IOP Conf. Series: Materials Science and Engineering **792** (2020) 012008 doi:10.1088/1757-899X/792/1/012008

is served by only one vehicle; Formula (6) ensure that the starting and ending points of distribution vehicles are in the distribution center; Equations (7) - (8) ensure that each colonel is served only once.

3. Genetic algorithm design

3.1. Genetic manipulation

3.1.1. Chromosome coding. In this paper, chromosome coding is carried out by natural number coding.

3.1.2. Fitness function. In the genetic algorithm, the higher the fitness $\boldsymbol{f}$, the better the chromosome, the smaller the total cost objective function $\boldsymbol{Z}$. $\boldsymbol{Z_i}$ will represent the total cost of the $\boldsymbol{i}$-th subpath. $\boldsymbol{f(i)}$ $=1/\boldsymbol{Z_i}, \boldsymbol{i = 1, \dots, N}$.

3.1.3. Select operation. The parent generation of the cross was selected by the roulette wheel selection strategy. At the same time, elite reservation mechanism is adopted based on roulette.

3.1.4. Cross operation. In this paper, the partial matching crossover rule is used to prevent the genetic algorithm from getting into the local optimum and the result from getting the optimal value.

4. Analysis of examples

4.1. Initial data

The data in [6] are adopted in this paper. The parameters in the model are set as follows: the running speed of the vehicle v=40km/h, the vehicle unit distribution cost is P_r=1.5, the fixed cost of each vehicle is f_k=85 , ε_1=50 , ε_2=100 , and the maximum load of the vehicle is 250 units. In the genetic algorithm, the population size is 150, the crossover probability is 0.9, the mutation probability is 0.05, and the maximum number of iterations is 200.

4.2. Analysis of results

The path obtained by using genetic algorithm in literature [6] is 0-7-13-14-15-0;0-11-9-1-6-10-0;0-12-3-5-4-8-2-0; As the path, the total cost is 499.8, the total path length is 163.2, and the average customer satisfaction is 0.813. We ran the optimal value of the objective function with a total cost of 476.3, a total path length of 141.5, and an average customer satisfaction of 0. 805. The vehicle routes of the three vehicles corresponding are shown in figure 2.

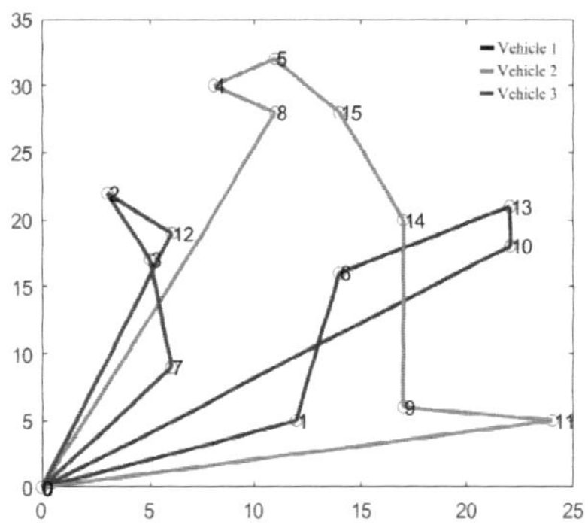

Figure 2. Vehicle routing.

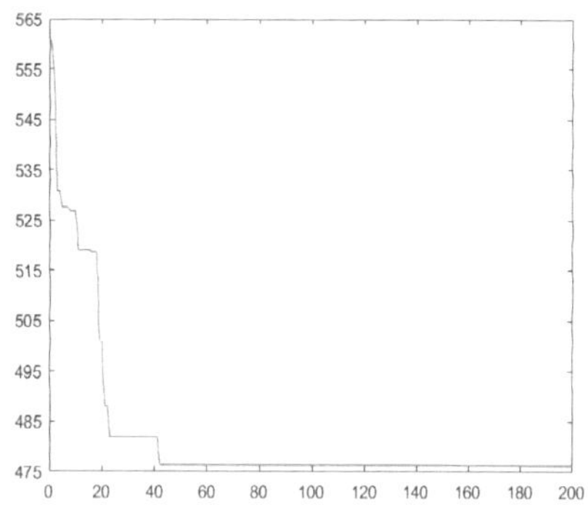

Figure 3. Optimization process.

It can be seen that when the experimental data and parameters are set exactly the same, the path length obtained by establishing the model in this paper is reduced by 15% compared with the results in literature [6], the average customer satisfaction is slightly lower, and the total cost is even lower, reduced by 5%.

5. Conclusion

In view of the problems existing in community group-buying logistics distribution, this paper, under the constraint of customer satisfaction function of time window, constructs a mathematical model of distribution cost for community group-buying logistics distribution, analyses in detail the fixed cost, transportation cost and time window penalty cost, and then designs a genetic algorithm to solve them. Through the example comparison, the community group purchase logistics distribution model obtains satisfactory solution and good satisfaction. This paper provides valuable reference for enterprises to improve user stickiness and customer service experience in the fierce competition.

References

[1] Baradaran, V., Shafaei A, Hosseinian A H. (2019) Stochastic vehicle routing problem with heterogeneous vehicles and multiple prioritized time windows: Mathematical modeling and solution approach. Computers & Industrial Engineering, 131: 187-199.
[2] Ghoseiri, K., Ghannadpour, S.F. (2010) Multi-objective vehicle routing problem with time windows using goal programming and genetic algorithm. Applied Soft Computing, 10(4): 1096-1107.
[3] Marinakis, Y., Marinaki, M., Migdalas, A. (2019) A multi-adaptive particle swarm optimization for the vehicle routing problem with time windows. Information Sciences, 481: 311-329.
[4] Yassen, E.T., Ayob, M., Nazri, M. Z. A. (2017) An adaptive hybrid algorithm for vehicle routing problems with time windows. Computers & Industrial Engineering, 113: 382-391.
[5] Zhang,Y.M., Li,Y.M., Liu,H.O.(2019) Research on VRP optimization of satisfaction constrained multi-model cold chain logistics. Statistics & Decision,35(04):176-181.
[6] Liu,T.Y. (2019) Research on the optimization of warehouse distribution mode of the third party platform of community group-buying under time window constraint. Zhengzhou University.

IOP Conf. Series: Materials Science and Engineering **792** (2020) 012009 doi:10.1088/1757-899X/792/1/012009

The Forward Fuselage Rigidity Design Based on Sensitivity Analysis of The Aircraft

Mingchun Wang[1], Haimeng Zhao[1*], Xin Hao[2], Shanlin Sun[1] and Ruilan Chang[1]

[1] Guangxi Colleges and Universities Key Laboratory of Unmanned Aerial Vehicle (UAV) Remote Sensing, Guilin University of Aerospace Technology, Guilin, Guangxi, 541004, China

[2] Mechanical Analysis department, Shenyang Aircraft Design&Research Institute, Shenyang, Liaoning, 110035, China

[*]Corresponding author's e-mail: zhaohaimeng@guat.edu.cn

Abstract. In this paper, structural characteristics of the forward fuselage of an aircraft were analyzed. Loadbearing characteristics of longitudinal members and transverse members of the forward fuselage were summarized. After analysis, longitudinal members were identified as main factors that affected structural deformation. Then, finite element analysis and sensitivity analysis methods were combined to determine accurate stiffness sensitivity parameters of the forward fuselage structure. Through analysis of the displacement sensitivity curve, the law of changes in stiffness of the forward fuselage with changes in sizes of longitudinal members was found out. Design requirements on stiffness of the forward fuselage were well satisfied by controlling parameters on purpose. Then, some experiments were designed to compare calculated displacements of the forward fuselage. Experiment results proved the accuracy of the proposed analysis method, allowing designers to get rid of their experience-based the forward fuselage stiffness design methods and providing a theoretical basis for the forward fuselage stiffness design. The proposed method could also provide some reference for structural design of new or retrofitted planes.

1. Introduction

Modern fighters are more and more complex to meet various functional requirement. The fuselage of an aircraft is generally designed to install various types of equipment to ensure reasonable space allocation and unaffected equipment performance. For the easy installation and maintenance of airborne equipment, the fuselage tends to have many large openings, with discontinuous crossbars. Therefore, the design of fuselage stiffness is a complicated problem. What factors have an impact on the stiffness of the fuselage? How do they affect the fuselage? In the past, these problems were generally solved through experience, which made the stiffness design unscientific and not rigorous. Besides, they would also lead to a mismatch between the material and the structural stiffness, resulting in excessive strength and structural weight.

Sensitivity analysis has been applied in many fields, such as designing variable selection and structural optimization design[1-11]. Through displacement sensitivity analysis, sensitive factors of the structural stiffness of fuselage can be accurately determined. In this case, the physical parameters of components with great impacts can be adjusted purposively, which makes the structural force more reasonable and finally achieves the match between the bearing capacity and structural stiffness.

Therefore, the purpose of reducing aircraft structure weight, saving materials, and improving aircraft performance can be reached.

2. Introduction to the displacement sensitivity analysis method

The structural equilibrium equation is as follows,

$$[K][u]=\{P\} \tag{1}$$

Where $[K]$ refers to the overall stiffness matrix of the structure, $[u]$ represents the nodal displacement, and $\{P\}$represents the nodal load. The differential equation of Equation (1) to the design variables is described as:

$$[K]\frac{\partial\{u\}}{\partial x_i}+\frac{\partial[K]}{\partial x_i}\{u\}=\frac{\partial\{P\}}{\partial x_i} \tag{2}$$

When $\{P\}$is independent of the design variables, it can be obtained according to Equation (2) that

$$[K]\frac{\partial\{u\}}{\partial x_i}=-\frac{\partial[K]}{\partial x_i}\{u\} \tag{3}$$

where $\frac{\partial\{u\}}{\partial x_i}$ is the sensitivity of the structural displacement to the design variables, namely, the displacement sensitivity.

$\frac{\partial[K]}{\partial x_i}$ can be calculated by calculating of each element and then assembling. The element stiffness matrix is presented by isoparametric element as:

$$[K]^e=\int_v[B]^T[D][B][J]d_v \tag{4}$$

where $[B]$ refers to the strain matrix, which is related to the structural parameter x_i, $[D]$ represents the physical stiffness matrix, $[J]$ represents the jacobian transformation matrix, and v is the volume.

$$\frac{\partial[K]^e}{\partial x_i}=\int_v\frac{\partial[B]^T}{\partial x_i}[D][B][J]d_v+\int_v[B]^T[D]\frac{\partial[B]}{\partial x_i}[J]d_v+\int_v[B]^T[D][B]\frac{\partial[J]}{\partial x_i}d_v \tag{5}$$

See matrixes $[B]$ and $[J]$ in the references[12-15].

3. Case analysis (taking the fuselage stiffness design of an aircraft as an example)

3.1. Determination of influencing factors

Due to the redesign of structures such as the equipment cabin, the fuselage stiffness of aircraft changes. Through analyzing the structural characteristics of the aircraft, it is found that the longitudinal components of the fuselage mainly include the floor, the longitudinal wall, the side plate, the skin, and the radome, while the transverse members are mainly composed of the fuselage frame. The transverse members, mainly used for maintaining the shape, have little effect on the Y-direction deformation of the aircraft. Therefore, the longitudinal components are the major factors influencing the structural deformation. This analysis aims to achieve the correspondence between the structural deformation and the designed reference values by scientifically changing the geometric parameters of these components. Figure 1 shows the structures of the floor, longitudinal wall, side plate, skin, and radome, and Table 1 indicates the specific dimensional changes.

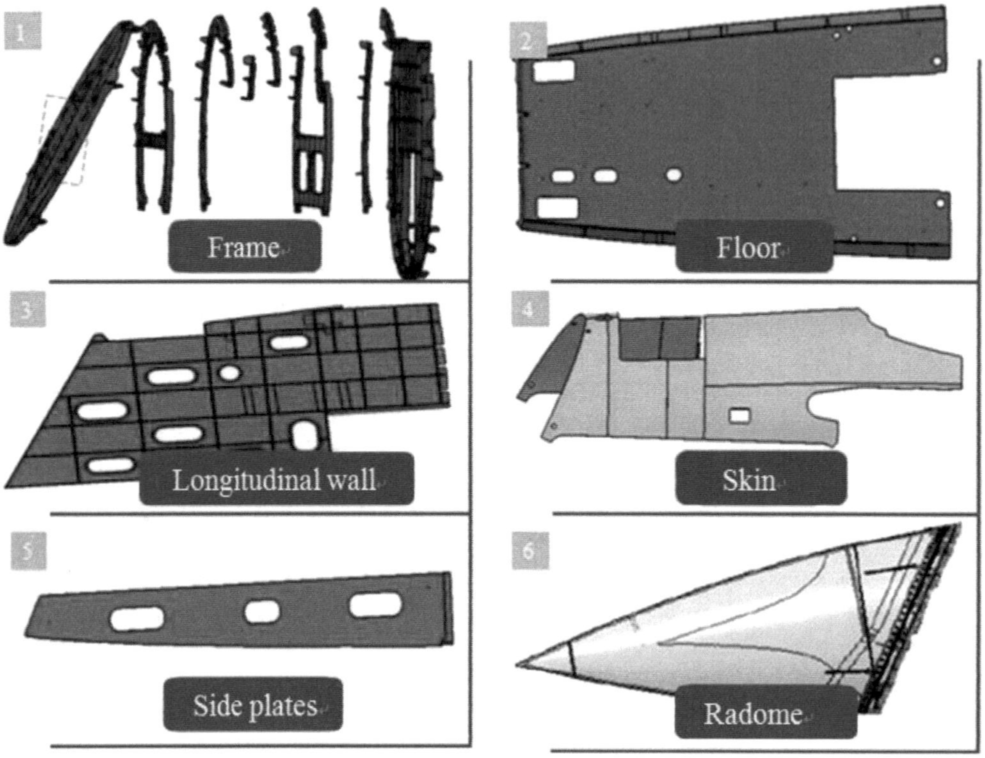

Figure 1. Structures of the influencing factors

Table 1. Thickness parameter ranges of the longitudinal components of the fuselage structures.

Number	Component	Thickness variation range (mm)
1	Upper and under floors	1.5～3
2	Longitudinal wall	1.5～3
3	Skin	0.8～1.5
4	Side plates	1.5～3
5	Radome	3～5

3.2. Displacement sensitivity analysis

The model shown in Figure 2 is modeled using Patran, and the displacement sensitivity analysis is performed on Point 1 and Point 2. Figure 3 shows the displacement change curves of the two points as the thickness parameters of the longitudinal components change. Figure 4 shows the sensitivity change curves of the two points as the thickness parameters of the longitudinal components change.

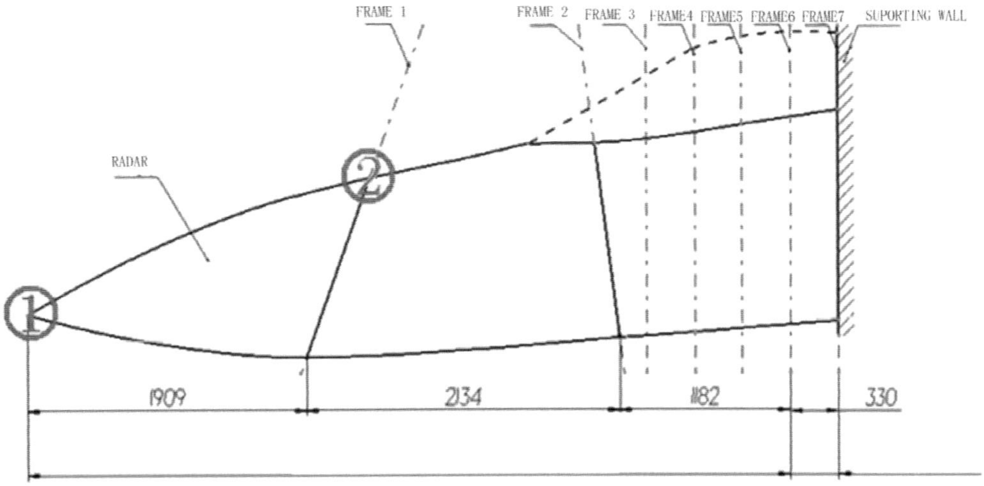

Figure 2. Fuselage structure and sensitivity analysis positions

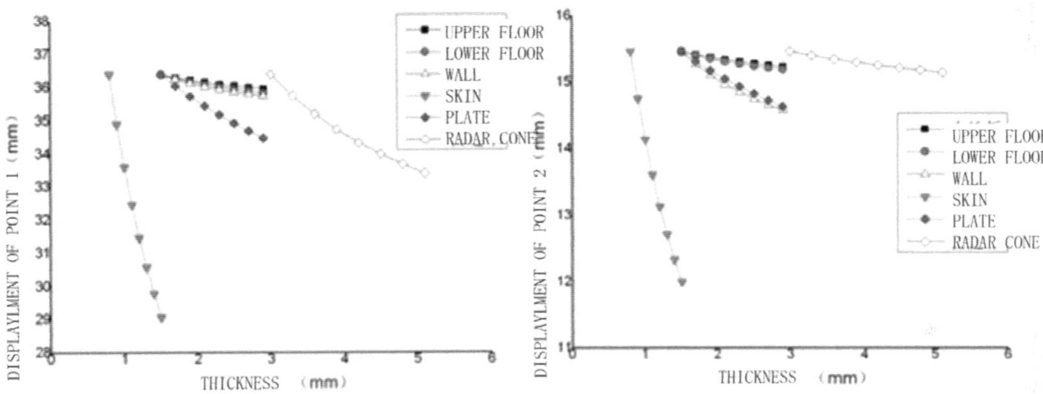

Figure 3. Displacement change curves of Point 1 and Point 2 as the thickness parameters of the longitudinal components change

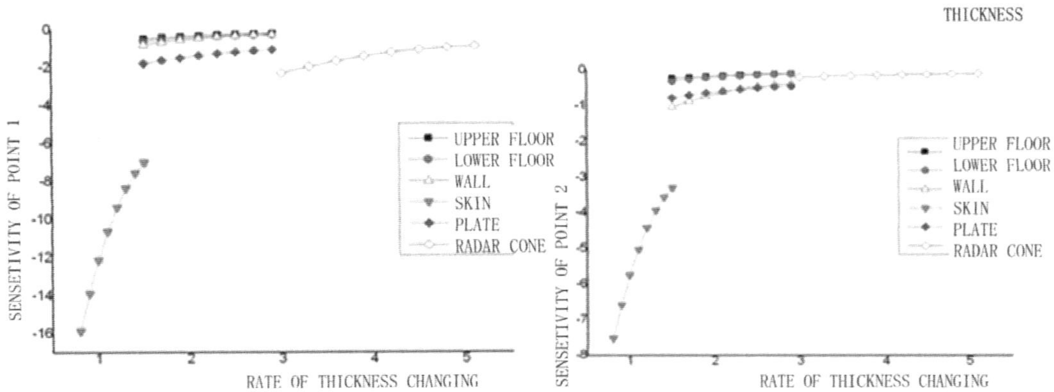

Figure 4. Sensitivity change curves of Point 1 and Point 2 as the thickness parameters of the longitudinal components change

Figure 3 suggests that the thickness variation of the skin has the most significant influence on the displacement of Point 1, followed by the thickness of the radome. According to the result of the displacement sensitivity analysis at point 1 in Fig. 4, Point 1 is most sensitive to the thickness of the skin. Besides, Point 1 is also sensitive to the thickness of the radome because there is only one longitudinal member, the radome, in front of Frame 1.

It can also be seen from Figure 3 that the thickness variation of the skin has the greatest influence on the displacement of Point 2, while the thickness variation of the other longitudinal members has little effect on it. According to the result of the displacement sensitivity analysis at point 2 in Fig. 4, Point 2 is most sensitive to the thickness of the skin. Since the radome is located in front of Frame 1, its thickness change does not affect Point 2. The displacement sensitivity analysis curve of Point 2 in Fig. 4 also proves that.

Through the above sensitivity analysis curve, it is clear that the skin thickness has the most considerable influence on the structural stiffness of Frame 4 of the fuselage. By purposefully controlling skin thickness and other parameters, the model can be adjusted to achieve the design target of fuselage stiffness. The comparison of the calculated displacement and design values of the fuselage of the aircraft is shown in Table 2 and Figure 5.

Table 2. Comparison of the computational displacement and design values of the fuselage of an aircraft (mm)

Location	Coordinate	Calculated value	Designed reference value	Calculated value / Designed reference value
1	774	-3.98	-4.16	0.96
2	1549	-2.68	-2.73	0.98
3	2349	-1.24	-1.34	0.93
4	2908	-0.74	-0.74	1.00
9	4420	0.00	0.00	-

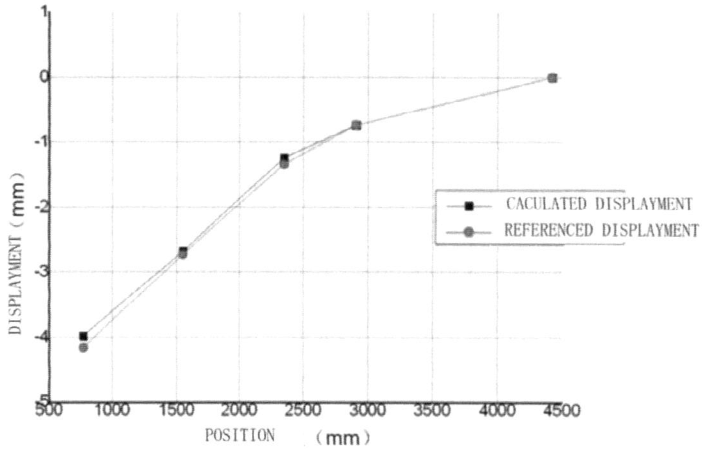

Figure 5. Comparison of calculated displacement and design values of the fuselage of an aircraft

4. Conclusion

Structural stiffness design is an indispensable part of the aircraft design. Due to the pursuit of combat performance, the weight control on modern aircraft is more and more strict. It is necessary to perform structural sensitivity analysis to get rid of dependence on human experience, scientifically and rationally select the physical parameters of the components, and match the bearing capacity with the structural stiffness, thus achieving structural optimization. In this paper, the finite element analysis technology and the sensitivity analysis method are fully utilized to determine the stiffness sensitivity parameters of the fuselage structure accurately. Through controlling the parameters purposefully, the design requirements of the fuselage stiffness of the aircraft are well achieved, which provides a reference for the design of the aircraft structure.

Acknowledgments

This work was supported by the key Research and Development Program of Guangxi (Grant No. AB18050014) and the National Natural Science Foundation of China (Grant No. 61841101).

References

[1] Keulen F V, Haftka R T, Kim N H. Review of options for structural design sensitivity analysis. Part1:Linear systems [J]. Computer Methods in Applied Mechanics and Engineering, 2005,194:3213～3243.
[2] Szymczak C. Sensitivity analysis of thin-walled members problems and applications[J]. Tin-Walled Structures,2003,41 :271～290.
[3] Zhen Luo, Liping Chen, Yuying Huang. Topology optimization design of continuum structure [J], Advances in Mechanics,2004,34(4):463—476.
[4] SIGMUND O'MAUTE K. Sensitivity filtering from a continuum mechanics perspective[J]. Structural and Multidisciplinary Optimization,2012,46(4) :471～475.
[5] Adelman H M, Haftka R T. Sensitivity analysis of discrete structural systems[J]. AIAA Journal, 1986, 24(5):823-832.
[6] Yangchen Deng, Weihong Zhang, Jihong Zhu. Problems arising from the structural design of thin-walled wing box [J], Mechanics in engineering, 2004,26(6):20—22.
[7] Papadrakakis M, Lagaros N D, Tsompanakis Y. Largescale structural optimization: computational methods and optimization algorithms[J]. Archives of computational Methods in Engineering, 2001, 8(3):239-301.
[8] Arora J S, Wang Q. Review of formulations for structural and mechanical system optimization[J]. Structural and Multidisciplinary Optimization, 2005,30(4):251-272.
[9] Yangchen Deng, Weihong Zhang, Jihong Zhu. Application of topology optimization method based on sensitivity threshold in aircraft stiffener design [J], Mechanics in engineering, 2005,27(4):26—31.
[10] Sobieski J. Structural Optimization by multilevel decomposition[J]. AIAA Journal, 1983, 23:1775-1782.
[11] Barthelemy M J. Engineering application of hierarchic multilevel optimization method. NASA Langleg Research Center. 1990. 1030-1037.
[12] Foundation of Optimum Design of Engineering Structure. [M]. Beijing：China Water & Power Press，1984.
[13] Principle and Application of Finite Element Method [M]. Beijing：China Water & Power Press，2009.
[14] Basic Principle of Finite Element Method and Numerical Method [M]. Beijing：Tsinghua University Press，2001.
[15] Matrix Structure Analysis Theory [M]. Beijing：National Defense Industry Press，1974.

An accurate and fast detection method for railway lightning protection component

Suining Wu[1*], Bin Li[1], Zhixin Wang[1], Xiaoguang Chen[1], Long Shi [1] and Shulin Tan[1]

[1] Institute of Intellective Technology, China Railway Signal & Communication Research & Design Institute Group Co., Ltd., Beijing, 100070, China

[*]Corresponding author's e-mail: wusuining@crscd.com.cn

Abstract. Lightning protection component is used to display the working condition of railway cable networks. Lightning strike damage cable networks, hence, the research of detection method for railway lightning protection is of great importance. So far, there are limited methods to solve this problem. Existing methods are based on electrical and magnetic signals. However, these method can not meet the acquirement of speed and accuracy. In this paper, we proposed a accurate and fast detection method for railway lightning protection component. The proposed method is based on deep neural network named W-YOLO. To solve the imbalanced samples problem of fault components and common components, we devised a novel classification loss function according to the proportion components. Since the raw class and location coordinate of components are lack of meaning, a sorting method of elements based on the angel and distance threshold is proposed. From the experiments, the proposed detection method is able to detect railway lightning protection component accurately in real time.

1. Introduction

Lighting strike leads greatly danger for railway cable network, which is harm for the safety of passengers and freight. At present, non-visual methods are mainly used to detect lightning protection components, such as electrical and magnetic signals[1-2]. Comparing with traditional manual detection method and non-vision detection methods, vision detection methods have the advantage of non-contact, high accuracy, dynamic and so on.

With the development of deep learning, lots of modern object detection methods based on convolutional neural networks(CNN) are proposed. Compared with traditional image process methods, they make detection accuracy and speed improved a lot. Thereby, we proposed an accurate and fast detection method for railway lighting protection component based on CNN.

In fact, the most useful results are the index of component other than the raw coordinates. Hence, we divide the detection task into two sub-task: component detection and component sort. Based on the analysis above, the following works are accomplished in this paper. First, we devised a detection framework based on CNN. To improve the detection ability for fault components, a novel classification loss function is proposed. In the last, we proposed a sort method to determine the unique index of component.

The rest of this paper is organized as follows. Section 2 introduces the related works about object detection. Section 3 introduces the proposed detection method for railway lighting protection component in detail. In section 4, experiments are conducted to illustrate the accuracy and feasibility of the proposed method. Conclusions are drawn in section 5.

IOP Conf. Series: Materials Science and Engineering **792** (2020) 012010 doi:10.1088/1757-899X/792/1/012010

2. Related works

2.1. Detection method based on low-level image pattern
Before the largely application of CNN, classic artificial designed features are most popular to realize object detection, such as Haar-like features, histogram of oriented gradient features and so on. Viola-Jones[3] proposed a face detection method based on Harr feature. HOG based methods are firstly proposed in pedestrian or general object detection, such as the HOG[4] and DPM[5]. These methods are limited in assigned tasks, that is, the performance drops when they are adopted in other tasks. What's more, they work very slow. Further works are conducted to improve the above methods. However, they are still not comparable to methods based on CNN.

2.2. Detection method based on CNN
Related works of CNN for object detection can be divided into two kinds: one-stage methods and two-stage methods. Two-stage methods inference the proposal boxes at first, then the following networks inference the classification by the pooled feature map. R-CNN series[6-8] are the most successful two-stage object detection networks. Thereafter, lots of methods further improve the speed and precision of R-CNN series. However, the speed and parameter quantity cannot meet the firm requirement in industry.

Unlike two-stage methods, one-stage methods do not rely on proposal regions. Detection time of one-stage methods is significantly less than two-stage methods, meanwhile, the accuracy of one-stage methods is sufficient for practical application. Naturally, one-stage methods are widely used in industry, especially automatic driving, face recognition and so on.

SSD [9-10]and YOLO families[11-13] are most successful one-stage object detectors. They are able to realize the fast and accurate object detection at the same time. YOLOv3 is the latest and best method combining the FPN and multi scale detection.

3. Alogrithm
The end-to-end YOLO detectors avoid using region proposal methods. YOLO detectors utilized raw images as the input and output the bounding boxes and classification of object. The detection time of YOLO detectors is significantly less than two-stage detectors. YOLO detectors perform fast and accurate at the same time. Moreover, several tiny-YOLO networks are proposed to further decrease parameters and improve speed.

Previous versions of YOLO predict on the highest level feature maps, which make them can achieve relatively high detection accuracy. However, localization errors occurs when YOLO performs classification and localization simultaneougsly.YOLOv1 and YOLOv2 inference the class and bounding boxes via the highest feature maps. However, the last convolutional layer is often spatially course, which is not sufficient for localization. Inspired by FPN[14], residual network and multi-feature map detection are used in YOLOv3 to improve the detection accuracy. Especially, the detection accuracy for small and medium size objects is improved by YOLOv3.

Another challenge to improve YOLO detectors is foreground-background class imbalance at training. He et al. [15] devised focal loss to solve foreground-background class imbalance problem. With this function, detection network focus on more difficult negative examples. However, Redmon report that focal loss cannot improve YOLOv3. Derakhshani et al. [16]proposed assisted excitation method to help the detection network train better with the manually excited activations.

3.1. Alogrithm
The proposed detection framework is illustrated in figure 1. W-YOLO utilize YOLOv3 as the backbone. As shown in figure 1, W-YOLO simultaneously inference the object class and location according to three different scale feature maps. Thereafter, W-YOLO achieves highly precision in object detection. YOLOv1 and YOLOv2 are limited to detect small size object. Combining the upsampled feature maps from high level, the sematic information is improved. As a results, W-YOLO is able to improve the detection for railway lightning protection component. Non maximum suppression is used to delete

duplicated and inaccurate results. At last, the accurate bounding box coordinates are used to determine the index of each component.

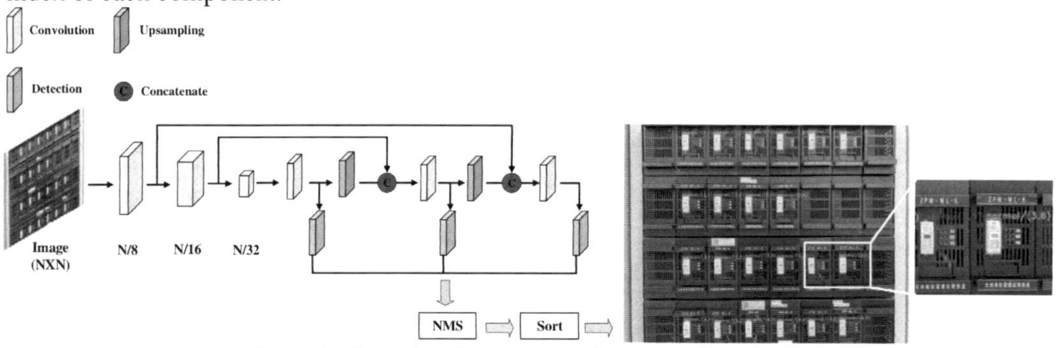

Figure 1. Detection framework of proposed method

3.2. *Loss function*

The original total loss of YOLOv3 can be divided into: classification loss, location loss and confidence loss. In W-YOLO, location loss and confidence loss remain the same.

We find that failure components are extremely less than common components in real images. If we directly train the network with raw dataset, the network will tend to detect an object as failure. In practical, the most important demand of our network is to detect failure components accurately. In the original loss function of YOLOv3, classification losses of two kinds of components contribute equally as shown in equation (1).

$$L_{class} = L_{class}^{N} + L_{class}^{F} \tag{1}$$

In this paper, inverse proportion of the quantity is utilized as the weight to adjust the loss dynamically. Thereafter, the network will learn to predict two kinds of components more equitable. The modified loss is established in equation (2).

$$L_{class} = \frac{n_F}{n_N} L_{class}^{N} + \frac{n_N}{n_F} L_{class}^{F} \tag{2}$$

3.3. *Sort method*

After the objects are detected, the bounding box coordinates are used to sort all the components. Figure 2 shows sorting process of lighting protection components. Firstly, we calculate the center point coordinates of each bounding box. After sorting process is completed in row and column, the unique index of lighting protection element is determined.

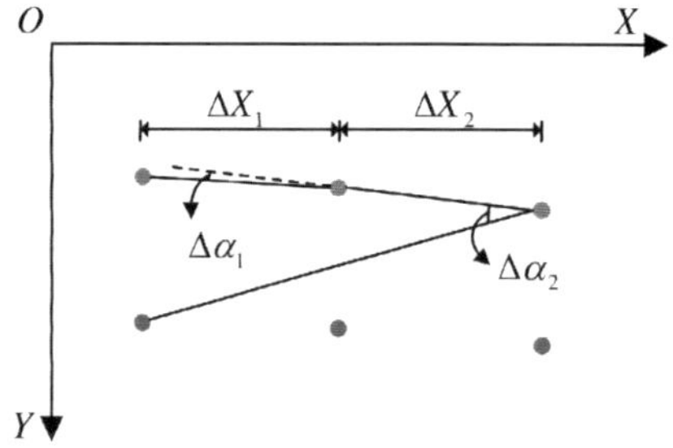

Figure 2. Sorting process of lighting protection components.

First, the X coordinates of center points are sorted in ascending order. The difference value of coordinates are calculated. Two adjacent points define a line. For the adjacent lines, the difference value of angels are calculated. In this paper, components will be located to the next column index while both the difference value of coordinates and angels larger than threshold.

Similarly, the row index can be determined by the Y coordinates of center points. Hence, we can obtain the unique index of each component.

4. Experiments

To evaluate the efficiency of proposed method, we collect a LPC dataset which containing 300 images of lighting protection components. During the training process, data augmentation is adopted. We train the proposed detector with 8 GPUs for 50 epochs. The initial learning rate is 0.001, and it is decreased by 0.1 after 25 epochs.

Figure 3 shows the detect results via our method. Figure 3(a) and figure 3(b) are origin images captured by camera, figure 3(c) and figure 3(d) are rotated images. Results show that the proposed method is able to process rotated images and origin images accurately. Thereafter, the proposed method is accurate and stable in practical application.

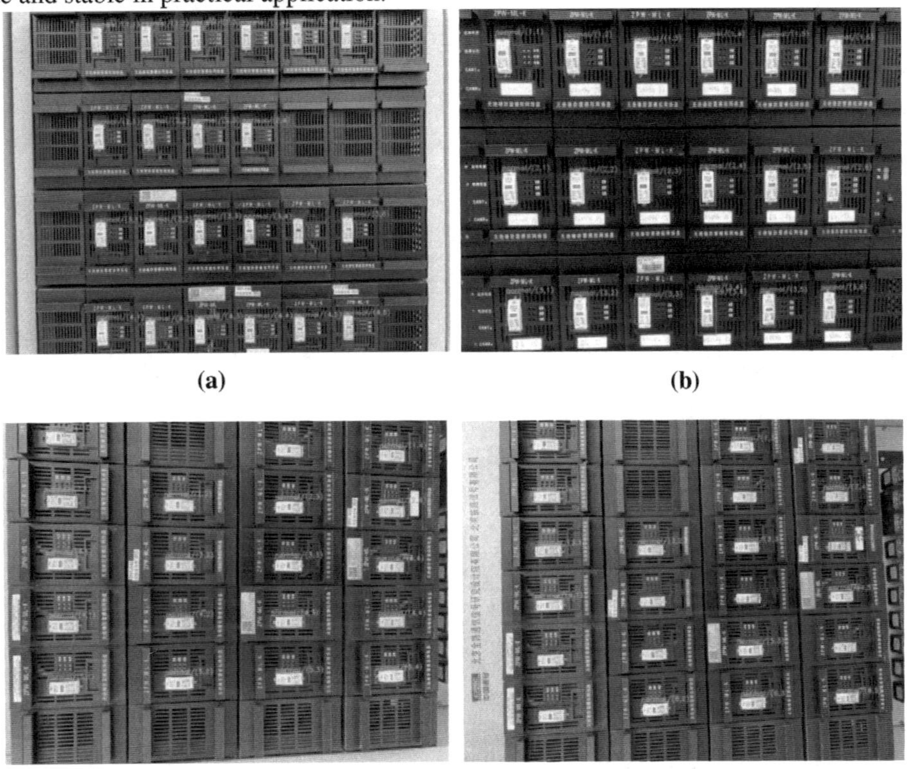

(a) **(b)**

(c) **(d)**

Figure 3. Detect results via proposed method. (a) and (b) are origin images; (c) and (d) are rotated images.

In table 1, we compare W-YOLO with YOLOv3 on LPC dataset. After the modification of classification loss, mean average precision(mAP) is increased by 2.7%.

Table 1. Comparison between W-YOLO and YOLOv3

Method	mAP
YOLOv3	89.4
W-YOLO	**92.1**

Detail AP comparison for each class is established in table 2. From the table, AP of each class is increased via the proposed method. What's more, AP difference between two class is decrease. That is, W-YOLO is able to solve label uneven problem.

Table 2. AP Comparison between each class

Method	Normal	Failure
YOLOv3	89.9	88.9
W-YOLO	**92.5**	**91.7**

Note that, we only modify the classification loss function to address the class label imbalance problem. The network architecture and detection process are not changed, Thereafter, the detection time is same with YOLOv3. For the proposed method, the detection time and sort time are 23ms and 2ms, respectively.

Figure 4 shows the detection results via our method and YOLOv3. Figure 4(a) and figure 4(b) are detection results via our method, figure 4(c) and figure 4(d) are detection results via YOLOv3. Missed detection problem happened in Figure 4(b) and figure 4(d). What's more, a failure component is detected to be common via YOLOv3. In practical, these mistakes make serious influence. It is worth noting that sort results are accurate though errors occur when YOLOv3 is adopted.

Figure 4. Comparison of detection results via our method and YOLOv3. (a) and (b) are detection results via our method; (c) and (d) are detection results via YOLOv3.

5. Conclusions

In this paper, we proposed a detection method to monitor the state of lighting protection components. We utilize a deep CNN to detect lighting protection components. YOLOv3 is adopted as the backbone, meanwhile, a modified classification loss based on the quantity of objects is proposed to address label imbalance problem. To evaluate the proposed method, we construct a dataset of lighting protection components. From the experiments, the accuracy and efficiency perform well in application.

References

[1] De Bruin, T., Verbert, K., Babuska, R. (2016) Railway Track Circuit Fault Diagnosis Using Recurrent Neural Networks. J. IEEE Transactions on Neural Networks and Learning Systems., 1-11.

[2] Verbert, K., Schutter, B. D., Babuska, R. (2015) Exploiting spatial and temporal dependencies to enhance fault diagnosis: Application to railway track circuits. Proceedings of the 2015 European Control Conference. Linz. pp. 3052–3057.

[3] Viola, P., Jones, M. J. (2004) Robust real-time face detection. J. International journal of computer vision., 57(2):137-154,

[4] Dalal, N., Triggs, B. (2005) Histograms of oriented gradients for human detection. IEEE Computer Society Conference on Computer Vision and Pattern Recognition . Piscataway. pp. 886-893.

[5] Yan, J., Lei, Z., Wen, L., Li, S. (2014) The fastest deformable part model for object detection. IEEE Conference on Computer Vision and Pattern Recognition. Columbus. pp. 2497-2504.

[6] Girshick, R., Donahue, J., Darrell, T., Malik, J. (2014) Rich feature hierarchies for accurate object detection and semantic segmentation. IEEE Conferenc on Computer Vision and Pattern Recognition.Columbus. pp. 580-587,

[7] Girshick, R. (2015) Fast R-CNN. IEEE International Conference on Computer Vision. Santiago. pp. 1440-1448

[8] Ren, S., He, K., Girshick, R., Sun, J. (2017) Faster R-CNN: towards real-time object detection with region proposal networks. J. IEEE Transactions on Pattern Analysis and Machine Intelligence., 39(6):1137-1149.

[9] Liu, W., Anguelov, D., Erhan, D., et al. (2016) SSD: Single Shot Multibox Detetor. European Conference on Computer Vision. Amsterdam. pp. 21-37.

[10] Fu, C. Y. , Liu, W. , Ranga, A. , Tyagi, A., Berg, A. C. (2017). DSSD : Deconvolutional Single Shot Detector. https://arxiv.org/abs/1701.06659.

[11] Redmon, J., Divvala, S., Girshick, R., Farhadi, A. (2016) You Only Look Once: Unified, Real-Time Object Detection. IEEE Conference on Computer Vision and Pattern Recognition. Piscataway. pp. 779-788.

[12] Redmon, J., Farhadi, A. (2016) YOLO9000: Better, Faster, Stronger. IEEE Conference on Computer Vision and Pattern Recognition. Honolulu. pp. 6517-6525.

[13] Redmon, J., Farhadi, A. (2018) YOLOv3: An Incremental Improvement. https://arxiv.org/abs/1804.02767.

[14] Lin, T. Y., Dollár, P., Girshick, R. , He, K., Hariharan, B., Belongie, S. (2017) Feature pyramid networks for object detection. IEEE Conference on Computer Vision and Pattern Recognition. Honolulu. pp. 936-944.

[15] Lin, T. Y. , Goyal, P., Girshick, R., He, K., Dollár, P. (2017) Focal loss for dense object detection. J. IEEE Transactions on Pattern Analysis & Machine Intelligence., 99: 2999-3007.

[16] Derakhshani, M. M., Masoudnia, S., Shaker, A. H., Mersa, O., Araabi, B. N. (2019) Assisted Excitation of Activations: A Learning Technique to Improve Object Detectors. https://arxiv.org/abs/1906.05388.

IOP Conf. Series: Materials Science and Engineering **792** (2020) 012011 doi:10.1088/1757-899X/792/1/012011

Research on Vehicle Lane Changing Characteristics of Multi-lane Type Highway Maintenance Operation Area Based on Vehicle Trajectory

Huang Yong*, Fang Daqing, Tan Fuliang, Tao Minglu, Si Daoguang, Shu Yang

Anhui Traffic Holding Group Co., Ltd., Hefei 230071, China

*Corresponding author's e-mail: 578798065@qq.com

Abstract: Aiming at the safety problem of vehicle lane change in highway maintenance operation area, a survey plan for vehicle lane change traffic data based on the combination of drone and video materials was developed. The vehicle trajectory was collected to analyze the maintenance work of highways in different lanes and vehicle lane change characteristics in the zone. The research shows that when the number of lanes increases, the lane change distance of the forced lane change on the closed lane increases, and the distance between the starting point and the working area increases. The closer to the closed lane, the higher the lane change probability of the vehicle, and the vehicle trajectory is concentrated. The probability of free lane change in the pass lane, the distance between the start point of the lane change and the work area and the lane change distance increases with the increase of number of lanes.

1. Introduction

As an important part of expressway maintenance, highway maintenance work areas affect the traffic capacity of expressways and increase the risk of driver transfer [1]. The main types of accidents in the highway maintenance work area are the same direction scraping, and the vehicle lane change is the main type of the same direction scraping accident [2-3]. The existing research on traffic safety in highway maintenance work area mostly focuses on vehicle driving characteristics and road traffic flow characteristics. Toledo [4] used NGSIIM data to estimate the parameters of the lane change regression model and made a significant analysis. Li [5] used the simulation method to study that the lane change duration increased and the maximum traffic volume decreased. Salvucci [6] used a driving simulator to simulate multiple lanes. Based on the vehicle trajectory analysis, this paper analyzes the influence of the number of lanes in the highway maintenance operation area on the vehicle lane changing characteristics based on the vehicle inspection traffic data in the highway management information system of Anhui Province, and provides the safety evaluation and accident for the highway maintenance operation area. The rate reduction provides theoretical support and also provides a data foundation for traffic simulation in the work area.

2. Vehicle lane change trajectory collection in multi-lane type maintenance work area

2.1. Data collection

This paper selects the road section of the highway maintenance operation area with two-way four-lane, six-lane and eight-lane closed inner lanes as the research object. In order to make the selected road sections universal, the following three rules are taken as the selection principle of the research road sections: the selected highway maintenance work area road sections have complete safety facilities and the facilities are arranged in accordance with the standards; the selected highway maintenance operation area road sections are all For flat road sections, there are no special linear features and slope characteristics; the selected highway maintenance work area sections are far away from the ramps and highway entrances and exits. The Hefei Ring Expressway was selected for research. The Hefei Ring Expressway has a good road condition with a designed speed of 120 kilometers per hour. It consists of two-way four-lane, six-lane, and eight-lane roads. The maintenance operation closes the lane (lane 1) on the inner side of the central separation zone. As shown in Figure 1, the construction sign will be displayed. The card is located at a distance of 1600 meters in the isolation zone of the upstream transition zone. This is the starting point of the warning zone. This article will focus on the section from the warning zone to the middle of the work zone in the highway maintenance operation area. This section is the accident in the highway maintenance operation area. Concentrated road sections, which can be used as research objects to make the research results clearer, avoid other low-risk road sections, and ensure the anti-interference of vehicle operating characteristics, and help to study the lane changing characteristics of the multi-lane type highway maintenance work area.

2.2. Data Processing

According to the vehicle trajectory, the vehicle lane changing characteristics are studied. The distance between the vehicle lane change and the lane line touch point and the lane change after the end of the lane change is finally determined as the lane change length L. Figure. 2 shows A schematic diagram of the position of the vehicle at one lane change, wherein the direction of travel of the vehicle is the direction of the abscissa, and the starting point of the upstream transition zone is the origin of the abscissa.

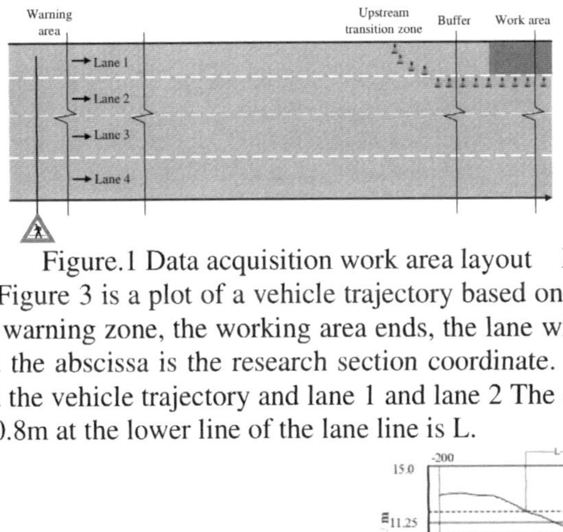

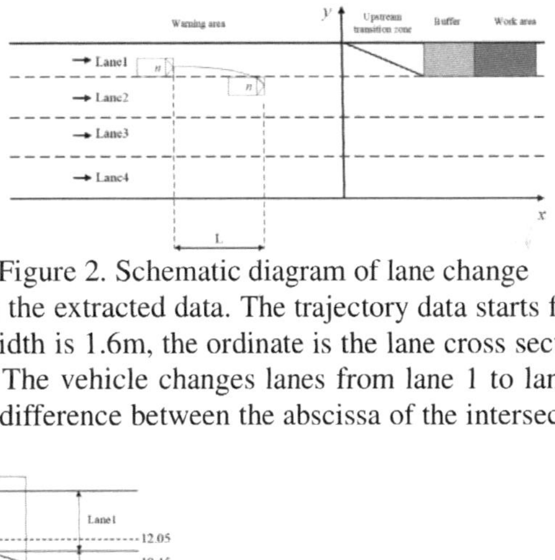

Figure.1 Data acquisition work area layout Figure 2. Schematic diagram of lane change

Figure 3 is a plot of a vehicle trajectory based on the extracted data. The trajectory data starts from the warning zone, the working area ends, the lane width is 1.6m, the ordinate is the lane cross section, and the abscissa is the research section coordinate. The vehicle changes lanes from lane 1 to lane 2, and the vehicle trajectory and lane 1 and lane 2 The difference between the abscissa of the intersection of 0.8m at the lower line of the lane line is L.

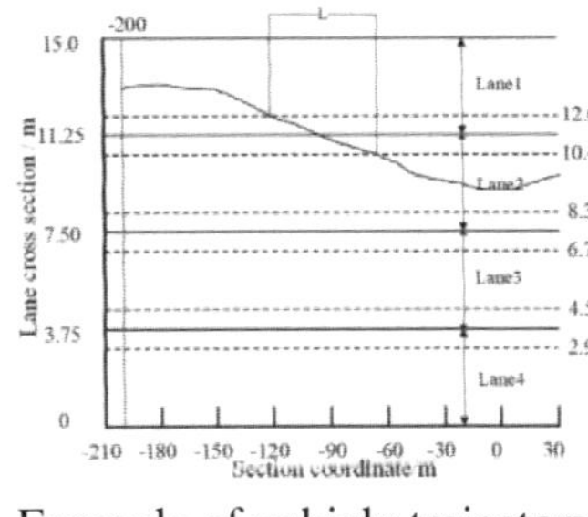

Figure 3. Example of vehicle trajectory extraction

3. Analysis of Vehicle Lane Change Characteristics in Maintenance Work Area

The vehicle lane change process is a comprehensive process and is the process of subjective decision-making by the driver based on external information. At present, the mainstream classification method of vehicle lane change can be divided into Mandatory Lane-Change, MLC and Discretionary

Lane-Change, DLC. The forced lane change in the highway maintenance work area refers to the forced lane change behavior performed by the driver due to the confluence of the maintenance work area and the reduction of the number of lanes. In contrast, the free lane change refers to the lane change behavior of the vehicle under the forced demand of the lane change, and is the lane change under the driver's complete subjective consciousness. After collecting the trajectory data in the video material, the 405 vehicle trajectory data of the multi-lane type maintenance work area including forced lane change and free lane change are selected, and Figure. 4 is the trajectory of each lane of the vehicle in the two-way four-lane maintenance work area. Figure 5 shows the running trajectory of each lane in the two-way six-lane maintenance work area. Figure 4 shows the trajectory of each lane in the two-way eight-lane maintenance work area.

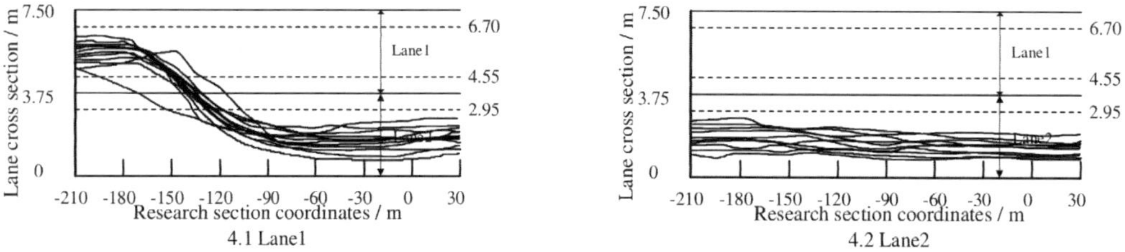

Figure 4. Vehicle trajectory of each lane in the four-lane maintenance work area

As shown in Figure. 4, in the two-way four-lane vehicle lane change type, the lane 1 vehicles all change lanes to lane 2, the vehicle lane change is all mandatory lane change, and the lane 2 vehicle has no lane change behavior. In Figure4.1, the initial point of lane change of lane 1 is completed at -130m on the abscissa, and the end point of lane change is completed at -105m on the abscissa. Most of the vehicles are concentrated on the abscissa -150m for lane change behavior. The lane change behavior is completed at the abscissa-120m, and the lane change length L is concentrated at 30m. In Figure4.2 , the initial ordinate distribution of the lane 2 is relatively scattered, and the rear axle distribution of the vehicle in the lane 1 is relatively concentrated.

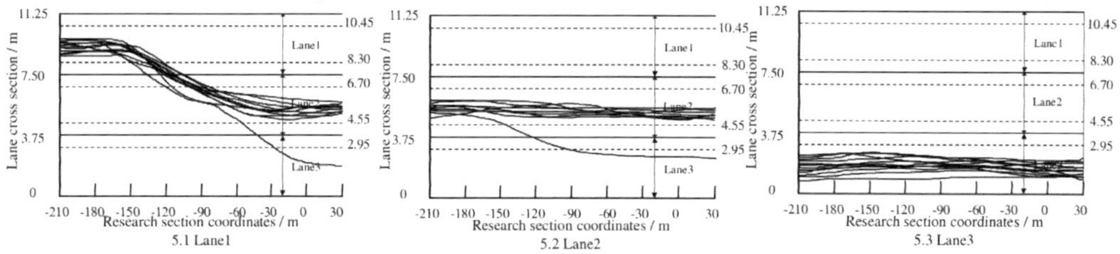

Figure 5.: Vehicle trajectory of each lane in the six-lane maintenance work area

As shown in Figure 5, in the two-way six-lane vehicle lane change type, lane 1 vehicles all perform lane change behavior, and a few have changed lanes twice to lane 2. The vehicle lane change is all mandatory lane change, and the second lane change is Free lane change, a small number of vehicles on lane 2 change lanes to lane 3, the lane change behavior is free lane change, and the vehicles on lane 3 do not change lanes. In Figure 5.1, the initial point of lane change of lane 1 is completed at -122m on the abscissa, the end point of one lane cha nge is completed at -90m on the abscissa, and the end point of the second lane change is completed at -27m, most of which are completed. The vehicle is concentrated on the abscissa -150m for lane changing behavior, and the lane changing behavior is concentrated on the abscissa -90m. The lane change length L is concentrated at 60m. In Figure 5.2, lane 2 receives the rear ordinate distribution of the vehicle in lane 1. More concentrated, some vehicles change lanes to lane 3, the lane change vehicle initial point is completed at -130m, the lane change end point is completed at -100m, In Figure 5.3, lane 3 receives lanes 1, 2 and the vehicle's rear ordinate distribution is concentrated.

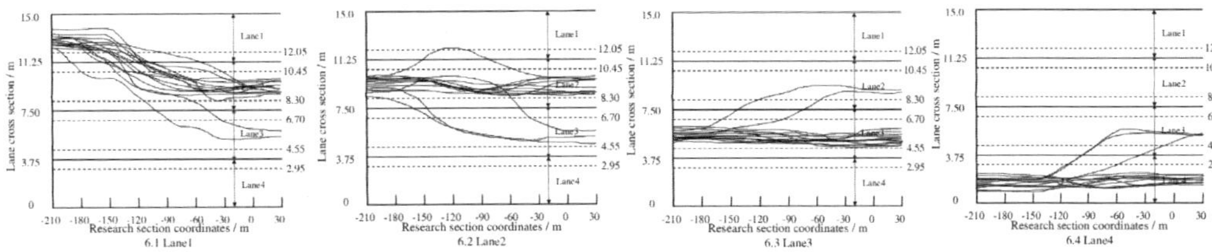

Figure 6.: Vehicle trajectory of each lane in the eight-lane maintenance work area

As shown in Figure 6, in the two-way eight-lane vehicle lane change type, lane 1 vehicles all perform lane change behavior, and a few have changed lanes to lane 3. The vehicle lane change is all mandatory lane change, and the second lane change is Free lane change, a small number of vehicles on lane 2 change lanes to lane 3, the lane change behavior is free lane change, and the vehicles on lane 3 do not change lanes. In Figure 6.1, the initial point of lane change of lane 1 is completed at -95m on the abscissa, the end point of one lane change is completed at -40m on the abscissa, and the end point of the second lane change is completed at -30m, most of which are completed. The vehicle is concentrated on the abscissa -145m for lane change behavior, focusing on the abscissa -90m to complete the lane change behavior, the lane change length L is concentrated at 55m, and the lane 2 in Figure 6.2 receives the rear ordinate distribution of the vehicle in lane 1. More concentrated, some vehicles change lanes to lane 3, the lane change vehicle initial point is completed at -150m, and the lane change end point is completed at -130m.In Figure 6.3, lane 3 receives lanes 1, 2 and the rear ordinate distribution of the vehicle is concentrated, and some lanes are lane-changed to lane 2. The lane change vehicle initial point is completed at -100m, and the lane change end point is at -60m.In Figure 4.4, part of the lane is changed to lane 3, the lane change vehicle initial point is completed at -110m, and the lane change end point is completed at -60m.

4. Conclusions

This paper analyzes the vehicle lane changing characteristics of highway maintenance work area based on vehicle trajectory. The conclusion shows that all the vehicles on the closed lane are forced to change lanes, and the vehicles in the remaining lanes are randomly changing lanes. The closer the vehicle is to the closed lane, the higher the lane change probability and the trajectory of the vehicle is concentrated. When the number of lanes increases, the lane change distance for the forced lane change of the vehicle on the closed lane increases, and the distance between the starting point and the work area increases. The end point of the forced lane change in the six-lane and eight-lane lanes is similar, and the forced lane change end point and the working area distance are reduced compared with the four lanes. When the number of lanes increases, the probability of free lane change of the passing lane increases, the lane changing distance increases, the distance between the starting point of the lane changing and the working zone increases, and the ending point of the lane changing is similar.

Acknowledgments

The project relies on the technology project of Anhui Transportation holding Group Co., Ltd.(Project No:GSKY-2019-01).

References:

[1] Chung Y , Kim H , Park M . (2012) Quantifying non-recurrent traffic congestion caused by freeway work zones using archived work zone and ITS traffic data. J . Transportmetrica, 8(4):307-320.
[2] Mohan, S. and P. Gautam. (2002) Cost of highway work zone injuries. J . Practice Periodical on Structural Design and Construction, 7(2): 68-73.
[3] Jiang X , Adeli H .(2010) Object-oriented model for freeway work zone capacity. J . Computer-Aided Civil and Infrastructure Engineering, 2010, 19(2):144-156.
[4] Toledo T., Zohar D. (2007) Modeling duration of lane changes . J . Transportation Research Record:

Journal of the Transportation Research Board, , 1999 (1): 71-78.
[5] Li X. G., Jia B., Jiang R. (2009) The Effect of Lane-Changing Time on the Dynamics of Traffic Flow . First International Conference on Complex Sciences, 2009, Shanghai, China,
[6] Salvucci D. D., Liu A. (2018) The time course of a lane change: Driver control and eye-movement behavior. J . Transportation Research Part F: Traffic Psychology &Behavior , 5(2):123-132.

Prediction of Road Traffic Accidents based on Rolling-optimized Grey Markov Mode

ChengzhenXu[1], HengZhang[1], YananGong[1], HuiyunSang[1*], GuanglinSun[2] and JingChen[3]

[1] Department of Logistics Engineering, Shandong Jiaotong University, Jinan, Shandong, 250300, China

[2] Road Traffic Safety Research Center of the Ministry of Public Security, Beijing, 100000, China

[3] School of Architecture and Rail Transit, Xi'an Vocational and Technical College, Xi'an, 710000, China

*Corresponding author's e-mail: hysang@139.com

Abstract. In order to solve the problem that the accuracy of the accident prediction is affected by the time-effectiveness of the forecast data, the paper introduces the rolling optimization strategy based on the original data of the road traffic accident in the road section, and further establishes the rolling optimization-grey Markov dynamic prediction model. Using the Markov chain theory to explore the transition law between different states, the development trend of road traffic accidents volume is predicted, and the prediction accuracy of the random time series is further improved. Case analysis proves that the method has better prediction accuracy and practicability over a certain period of time, which can provide reference for road traffic accident prediction analysis and traffic safety early warning

1. Introduction

At present, there are two limitations in road traffic accident prediction methods. The first is that the requirements for prediction data are high, the second is that the accuracy of mathematical models is strict. Due to the randomness, non-linearity, volatility and time-varying of initial data, road traffic accident prediction is still difficult. Currently, traffic accident prediction are mainly based on data-driven and statistical analysis, such as time series method[1], support vector machine method (SVM) [2], Markov method[3], etc. Time series method can describe the periodic law of data, but it is unrealistic to record and predict all the data, so it is difficult to establish training and learning models. SVM has high prediction accuracy, but it needs a lot of training data, which consumes lots of energy and is difficult to be applied in practice. Markov can reflect the periodic variation characteristics of the data with strong randomness, but its determination of the state set is difficult. Grey prediction *GM* (1, 1) only needs a small sample to establish the model[4], but *GM* (1,1) has a large long-term prediction residual, which cannot reveal the regularity of periodic changes of data.

Based on above analysis and relevant research, this paper proposes a method based on Rolling Optimization - Grey Markov which combines Markov method with *GM* (1, 1) prediction model. Based on rolling data prediction theory[5], the model continuously adds new prediction data and eliminates old data, finally obtains prediction data during the study period.

2. The construction of the accident prediction model

2.1. Establish GM (1,1) rolling optimization model

Let $x^{(0)} = \left[x^{(0)}(t_1), x^{(0)}(t_2), \cdots, x^{(0)}(t_n)\right]$ be the original time series, where t_i is the time factor . According to Sequence $\{x^{(0)}\}$, predicts the traffic accident data of period *m* after period t_n. Set the predicted time series to $\tilde{x}^{(0)} = \left[\hat{x}^{(0)}(t_{n+1}), \cdots, \hat{x}^{(0)}(t_{n+m})\right]$. The modeling mechanism is as follows:

Model usually sets the sequence$\{x^{(0)}\}$as a discrete time series, thus introducing a first order cumulative sequence $\{x^{(1)}\}$.The randomness of the original sequence is weakened to enhance the regularity of the time series. Let the first order cumulative sequence $x^{(1)}$ be:

$$x^{(1)} = \left[x^{(1)}(t_1), x^{(1)}(t_2), \cdots, x^{(1)}(t_n)\right] \tag{1}$$

where $x^{(1)} = \sum_{k=1}^{i} x^{(0)}(t_k), i = 1,2,3, \cdots, n$

Time series $x^{(1)}$ is the monotonically increasing function of *t*. Then a single-sequence first-orderlinear dynamic GM(1,1) model is established, and the corresponding differential equation is:

$$\frac{dx^{(1)}(t)}{dt} + ax^{(1)} = u^{(1)} \tag{2}$$

In the GM model, the initial value of the boundary is satisfied with $x^{(0)}(t_1) = x^{(1)}(t_1)$, from which the approximate function can be obtained:

$$\tilde{x}^{(1)}(t_i) = \left[x^{(0)}(t_i) - \frac{u}{a}\right] \cdot e^{-a(t-1)} + \frac{u}{a}, i = 1,2, \cdots, n \tag{3}$$

Among them, *a* and *u* are the undetermined coefficients of the initial statistical data of the accident, which represent the development coefficient and the grey control quantity in the modeling process.They can be calculated by the least square method:

$$\hat{a} = [x(A)^T x(A)]^{-1} \cdot x(A)^T \cdot Y_N \tag{4}$$

where $\hat{a} = [a, u]^T$

$$x(A) = \begin{bmatrix} -\frac{1}{2}\left[x^{(1)}(2) + x^{(1)}(1)\right] & 1 \\ -\frac{1}{2}\left[x^{(1)}(3) + x^{(1)}(2)\right] & 1 \\ \vdots & \vdots \\ -\frac{1}{2}\left[x^{(1)}(N) + x^{(1)}(N-1)\right] & 1 \end{bmatrix} \tag{5}$$

$$Y_N = \left[x^{(0)}(2), x^{(0)}(3), \cdots, x^{(0)}(N)\right]^T \tag{6}$$

Therefore, the formula (7) is obtained as:

$$\tilde{x}^{(1)}(t_i) = \left(x^{(0)}(1) - \frac{u}{a}\right) e^{-a(t_i-1)} + \frac{u}{a} \tag{7}$$

Based on the above steps, the time series function of predicted value of the traffic accident amount can be obtained by the regressive inverse operation:

$$\tilde{x}^{(0)}(t_i) = \tilde{x}^{(1)}(t_i) - \tilde{x}^{(1)}(t_{i-1}) \tag{8}$$

$$\tilde{x}^{(0)}(t_i) = (1 - e^a) \cdot \left(x^{(0)}(1) - \frac{u}{a}\right) e^{-a(t_i-1)} \tag{9}$$

(3) Residual prediction GM(1,1) model

$\tilde{x}^{(0)}(t_i)$ is the factor prediction value of t_i, and then the residual sequence is constructed as:

$$\varepsilon^{(0)}(t_i) = \left\{x^{(0)}(t_i) - \tilde{x}^{(0)}(t_i)\right\} \tag{10}$$

IOP Conf. Series: Materials Science and Engineering **792** (2020) 012012 doi:10.1088/1757-899X/792/1/012012

The sequence $\{\hat{\varepsilon}^{(0)}(t_i)\}$can be found by the GM(1,1) model, which is similar to theformulas (2)-(10). Adds$\hat{\varepsilon}^{(0)}(t_i)$ of relevant sample to the original prediction sample $\tilde{x}^{(0)}(t_i)$ and anchieve the revised prediction data$\hat{x}^{(0)}(t_i)$:

$$\hat{x}^{(0)}(t_i) = \tilde{x}^{(0)}(t_i) + \hat{\varepsilon}^{(0)}(t_i) \tag{11}$$

2.2. Markov Residual Modified GM (1,1) Model

(1) Markov residual correction

The Markov method is based on state transition prediction, and the residual sequence is divided into h states. The residual value can only be expressed as positive and negative: $E1$ and $E2$, So h=2. Set the state transition probability of state E_i at time t_i to state E_j at time t_j after k steps of transition is $p_{ij}(k)$, Markov's state transition matrix $p_{ij}(k)$ can be calculated and obtained as:

$$P(k) = [p_{ij}(k)]_{2\times 2} \quad i,j = 1,2 \tag{12}$$

Defines the first step state transition probability vector $p(1) = [p_1 \quad p_2]$ (13)
where, p_1 is the probability of state E1, and p_2 is the probability of state E2.

After the state is transferred by k steps,we can get:

$$p^{(k)}(1) = [p_1{}^{(k)} \quad p_2{}^{(k)}] = p(1) \cdot P(k) \tag{14}$$

The Markov residual correction result is shown in equation (15):

$$\hat{x}_1^{(0)}(t_i) = \tilde{x}_1^{(0)}(t_i) + s(t_i) \cdot \hat{\varepsilon}^{(0)}(t_i) \tag{15}$$

In the formula, $\hat{x}_1^{(0)}(1) = \tilde{x}_1^{(0)}(1)$, and $s(t_i)$ is the residual symbol.

In step k, when $p_1 > p_2$, let the residual symbol $s(t_i) = 1$; and when $p_1 < p_2$, let the residual symbol $s(t_i) = -1$.

2.3. Rolling Optimization Strategy

Considering that the prediction accuracy is greatly affected by the timeliness of the data, this paper uses "rolling data prediction principle" to remove the first period of data $\tilde{x}^{(0)}(t_{n+1})$ of the original sequence and add the calculated new predicted value $\hat{x}^{(0)}(t_{n+1})$, re-construct the time series in the subsequent prediction process and then to predict $\hat{x}^{(0)}(t_{n+m+1})$[11,12] of the next period. The process circulates and can calculate a series of new forecast data, which is drawn as Fig.1.

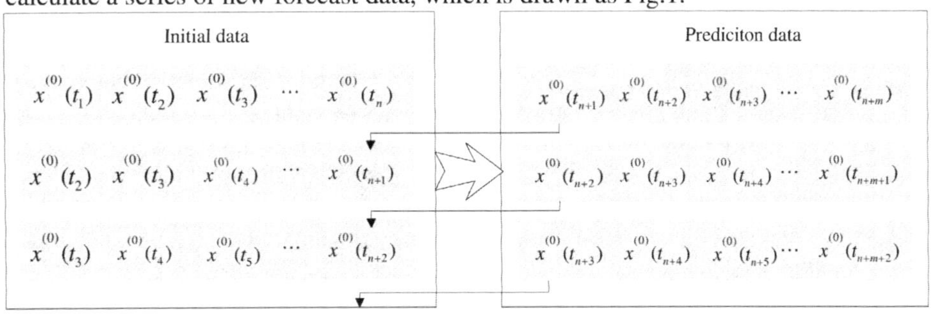

Figure1. The layout of rolling optimization strategy

3. Accuracy test of combined forecasting model

Let the residuals of the original time series and the predicted series be $\delta(t_i)$, then the mean value$\overline{\delta}(t_i)$ and variance $S_\delta$are as follows.

$$\overline{\delta}(t_i) = \frac{1}{n}\sum_{i=1}^{n} \delta(t_i), \quad S_\delta^2 = \frac{1}{n}\sum_{i=1}^{n}[\delta(t_i) - \overline{\delta}(t_i)]^2 \tag{16}$$

Mean value$\overline{x}(t_i)$and variance $S_o$of the original time series are:

$$\overline{x}(t_i) = \frac{1}{n}\sum_{i=1}^{n} x^{(0)}(t_i), \quad S_o^2 = \frac{1}{n}\sum_{i=1}^{n}\left[x^{(0)}(t_i) - \overline{x}(t_i)\right]^2 \tag{17}$$

Let the mean square error ratio C be:

$$C = \frac{S_\delta}{S_o} \tag{18}$$

Table 1.grade of accuracy inspection

Accuracy grade	Grade 1 (excellent)	Grade 2 (good)	Grade 3(qualified)	Grade 4(unqualified)
Mean square error ratio C	≤0.35	>0.35~0.50	>0.50~0.65	>0.65~0.80

4. Case study

(1) Predictive value calculation

Based on the statistical data of traffic accidents in 9 statistical periods of a subordinate administrative district of Jinan, the rolling optimization-grey Markov method is used to predict the traffic accidents in this district in the next few years. primary number of traffic accidents is shown in Tab 2.

Table 2.The primary number of traffic accidents

Statistical period	1	2	3	4	5	6	7	8	9
Accidents No .(hundred)	0.62	0.87	1.02	0.8	1.8	1.6	1.66	1.35	1.76

Based on the rolling optimization strategy, the first original data is sequentially removed, and the new calculated prediction value is added. According to Grey Markov theory, using Matlab to realize the algorithm. The results are achieved as in Tab. 3. And the value of C is 0.1675, i.e. $C<0.35$ showing the excellent prediction accuracy. It is proved that road traffic accident prediction model is effective and can be used for subsequent prediction to reflect the development trend of road traffic accidents.

Table3.The predicted value based on rolling optimization - Grey Markov

Statistical period	1	2	3	4	5	6	7	8	9
Results(hundreds)	0.62	0.79	0.99	0.77	1.79	1.49	1.67	1.49	1. 84

5. Conclusion

According to the characteristics of road traffic accidents and time series data, it is guided by the theory of gray system and Markov chain theory, a rolling optimization-grey Markov dynamic pr ediction model is established for road traffic accident volume prediction. The main contributions of this paper are as follows:

(1) The rolling optimization strategy has better improved the problem that the prediction accuracyisgreatlyaffectedbythetimelinessofthedata.Eachpredictionreplacesthefirstperioddata in the time series. Moreover, the method is in line with the actual situation, it is relatively simple and easy tooperate.

(2) The method provided in this paper is to predict the number of traffic accidents. The accuracy of the prediction results depends to a large extent on the quality and quantity of the observed samples. However, for the case of limited data, based on the rolling optimization strategy, a small amount of data can be used to complete the forecast, and case studies show that higher prediction accuracy can be achieved.

IOP Conf. Series: Materials Science and Engineering **792** (2020) 012012 doi:10.1088/1757-899X/792/1/012012

Acknowledgments
Authors wishing to acknowledge the anonymous reviewers for their careful, insightful, and constructive comments that led to an improved version of this paper. This work is partially supported by Technical Research Plan of the Ministry of Public Security(2018JSYJB05).

References

[1] Li S,Cui Z,Zhao J. Research and Application of Dynamic Triple Exponential Smoothing Method Based on Matlab Adaptation—Taking the Prediction of National Road Traffic Accidents as an Example[J].Mathematics in Practice and Theory,2018,48(12): 169-177.
[2] Tang Z, Dong W, Wang Z. Research on SVR-based urban road scraping accident assessment model [j]. Journal of Transportation Engineering and Information, 2019, 17 (01): 1-4.
[3] Cao J,Shi S,Cao H, *et al*.Study on the road accident and traffic accident prediction and relationship of hazardous chemicals based on GM(1,1)- Markov[J].China Safety Production Science and Technology , 2019, 15 (01): 26-31.
[4] Jiang S, Song K, Xie W, *et al*. Data Analysis of Road Traffic Accidents Based on Grey Correlation and Apriori Algorithm[J]. Highway Engineering, 2019, 44(04): 67-73.
[5] Xu G. Statistical Forecasting and Decision Making (Fifth Edition) [m]. Shanghai University of Finance and Economics Press.2016.

IOP Conf. Series: Materials Science and Engineering **792** (2020) 012013 doi:10.1088/1757-899X/792/1/012013

Research on location model of emergency material storage warehouse based on generalized maximum coverage

Zhong-hua Cheng[1,*], Qiang Wang[1,2,a], Min Du[1,b], Yong Chen[1,c]

[1]Department of Equipment Command and Administration, Army Engineering University, Shijiazhuang 050003, China

[2]Army Military Transportation University, Tianjin, 300161, China

*Corresponding author e-mail: 511091065@qq.com, [a]qw_up@foxmail.com, [b]17778891902@163.com, [c]675618842@qq.com

Abstract. It is of practical significance to analyze the support requirements of each rescue team for the rationality of warehouse location. In order to solve the problem that the traditional emergency material storage warehouse does not consider the difference of material guarantee timeliness of rescue team, a generalized covering location model is proposed. In this paper, the coverage is introduced to measure the quality of security services in each disaster area, and the maximum generalized coverage model is established to find the best location scheme of emergency material storage warehouse. Finally, the feasibility of the method is verified by an example, and the proposed method can better solve the location problem of emergency materials storage warehouse.

1. Introduction

When the rescue team supplies emergency materials, it is necessary to set up an emergency material storage warehouse to guarantee the emergency material supply in the disaster area. At present, the research methods of location problem are mainly divided into multi-attribute decision-making evaluation method [1] and four kinds of location optimization models (set coverage model, maximum coverage model, P median model, P center model) [2-5]. The above optimal location model is also widely used in fire stations, hospitals, military facilities and other emergency rescue location problems [6-8]. Based on the concept of "partial coverage" and "step-by-step coverage", the maximum generalized coverage model considers the influence of distance on the response degree of demand services, which can accurately reflect the degree of "coverage".

To sum up, a service satisfaction function of emergency material demand is constructed based on the location of emergency material reserve warehouse. The model considers the constraints of the maximum support delay time of each rescue area and the capacity constraints of the emergency material storage warehouse. Aiming at the problem of material supply in the rescue area which is not covered effectively, the penalty cost function of emergency material supply is constructed. The multi-objective maximum coverage location model is established, and the location of emergency material storage warehouse is solved. Finally, the feasibility of the model is verified by a case.

2. Problem Description

At the beginning of the disaster, the rescue team needs to arrange and select the location of the emergency material storage warehouse, so as to provide emergency material supply guarantee for each disaster area. The emergency material storage warehouse must be located in combination with the requirements of the disaster affected areas for material supply support. In this paper, through the construction of emergency material service satisfaction function, the optimization model of emergency material supply is established to maximize the overall satisfaction of emergency material support. Due to the complex disaster environment, in order to facilitate the analysis and research on the location of emergency materials storage warehouse, the following reasonable assumptions are made:

（1）Location information and material demand of each affected area are known;

（2）The maximum delay time of material supply in each affected area is known;

（3）According to the disaster situation, the number of emergency storage warehouse is known;

（4）Each emergency material storage warehouse has a capacity limit, and the upper limit of its inventory is known;

（5）To avoid using too many warehouse points, each disaster area is provided with an emergency material storage warehouse.

3. Modelling of Generalized Coverage Location

3.1. Satisfaction function of emergency material demand service

By constructing the service satisfaction function $F_i(t_{ij})$ of emergency material demand in each affected area, it reflects the satisfaction degree of emergency material demand response in the affected area $i(i=1,2,...,I)$ effectively covered to alternative emergency material reserve warehouse $j(j=1,2,...,J)$, $F_i(t_{ij}) \in [0,1]$.

$$F_i(t_{ij}) = \begin{cases} (1-\frac{t_{ij}}{T_i})^{\alpha_i} & 0 \le t_{ij} \le T_i \\ 0 & t_{ij} > T_i \end{cases} \quad (0<\alpha_i<1) \tag{1}$$

Where, t_{ij} is the time required for alternative emergency material storage warehouse j to reach disaster area i. T_i is the maximum guarantee delay time in the affected area. When the supply time of emergency materials exceeds T_i, the service satisfaction of emergency materials demand is 0. α_i is the time sensitivity of the affected area [9-10]. According to the time limit requirements for emergency supplies in the disaster affected area, $0<\alpha_i<1$ is taken here.

To sum up, the overall satisfaction of emergency materials demand and service in the disaster affected areas is as follows:

$$Z_1 = \sum_{i\in I}\sum_{j\in J} \omega_i d_i F_i(t_{ij}) x_{ij}^1 \tag{2}$$

Where, ω_i is the importance weight of each affected area. d_i is the emergency material demand of each affected area. x_{ij}^1 represents the decision variable that the emergency material storage warehouse can provide emergency material supply service to the disaster area, and meets the following requirements:

$$x_{ij}^1 = \begin{cases} 1 & \text{Emergency warehouse alternative } i \text{ provides material support for disaster area } j \\ 0 & other \end{cases} \tag{3}$$

3.2. Penalty cost function of emergency supply

According to the requirements of emergency material supply guarantee, the alternative points of emergency material storage warehouse can't effectively cover all the loading units. According to the principle of proximity, this paper provides emergency supplies to the affected areas that are not effectively covered by the emergency supplies storage warehouse with sufficient surplus stock and the closest distance. If the remaining inventory is not satisfied, it will be directly supplied by the general emergency materials warehouse. Therefore, the penalty cost function of emergency supplies is introduced to measure the penalty cost of emergency supplies, as follows:

$$P_{i'}(t_{i'j}) = \begin{cases} \gamma\left(\dfrac{t_{i'j} - T_{i'}}{\max(t_{i'j}) - T_{i'}}\right)^{\beta_{i'}} & t_{i'j} > T_{i'} \quad (1<\beta_{i'}) \\ 0 & 0 \le t_{i'j} \le T_{i'} \end{cases} \tag{4}$$

The penalty cost function $P_{i'}(t_{i'j})$ of emergency material supply reflects the penalty cost of alternative emergency material storage warehouse $j(j=1,2,...,J)$ supplied by disaster area $i'(i=1,2,...,I')$ not effectively covered. $P_{i'}(t_{i'j}) \in [0,1]$. $\beta_{i'}$ is the time sensitivity of the affected area. Since the penalty function reflects the degree of penalty for delay, $1<\beta_{i'}$. γ is the delay penalty factor, which is used to adjust the degree of delay penalty. $t_{i'j}$ represents the time from the affected area not effectively covered to the alternative points of each warehouse and the rear warehouse.

To sum up, the total penalty cost of emergency supplies in the disaster area is:

$$Z_2 = \sum_{i' \in I}\sum_{j \in J} \omega_{i'} c_{i'} d_{i'} P_{i'}(t_{i'j}) x_{i'j}^2 \tag{5}$$

Where, $c_{i'}$ is the penalty cost of emergency supplies per unit time. $\omega_{i'}$ is the importance weight of each affected area. $d_{i'}$ is the emergency material demand of each affected area. $x_{i'j}^2$ represents the emergency supplies storage warehouse that meets the supply requirements is selected from the remaining warehouses to provide emergency supplies to the affected areas that are not covered effectively, as follows:

$$x_{i'j'}^2 = \begin{cases} 1 & \text{Emergency warehouse alternative } i' \text{ provides material support for disaster area } j' \\ 0 & other \end{cases} \tag{6}$$

3.3. The maximum generalized coverage location selection model based on multi-objective constraints

To sum up, the generalized maximum coverage location model P_1 with multi-objective constraints is established:

$$\max Z_1 = \sum_{i \in I}\sum_{j \in J} \omega_i d_i F_i(t_{ij}) x_{ij}^1 \tag{7}$$

$$\min Z_2 = \sum_{i' \in I}\sum_{j \in J} \omega_{i'} c_{i'} d_{i'} P_{i'}(t_{i'j}) x_{i'j}^2 \tag{8}$$

Subject to:

$$\sum_{j \in J} y_j = p \tag{9}$$

$$\sum_{j \in J} x_{ij} = 1; \sum_{j \in J} x_{i'j} = 1 \tag{10}$$

$$\sum_{i \in I}\sum_{j \in J} d_i x_{ij} \le S_j \tag{11}$$

$$t_{ij} = \frac{L_{ij}}{a} \tag{12}$$

$$x_{ij}^1 \in \{0,1\}, x_{i'j}^2 \in \{0,1\}, \forall i \in I, j \in J, \ \forall i' \in I' \tag{13}$$

$$y_j \in \{0,1\}, j \in J \tag{14}$$

The model has two objective functions and belongs to multi-objective optimization model. In this paper, the linear weighting method is used to standardize multiple decision-making objectives, and the

IOP Conf. Series: Materials Science and Engineering **792** (2020) 012013 doi:10.1088/1757-899X/792/1/012013

location model P_1 is transformed into a single objective location model P_2 by using the model solution idea of reference [11].

$$Z_3=\max(k_1Z_1^*-k_2Z_2^*) \tag{15}$$

Subject to: Formula (9) ~ Formula (13).

Where, Z_1^* and Z_2^* are the decision objectives after standardization.

4. Case Study and Analysis

According to the analysis of the situation of the disaster area, it is urgent to supply emergency materials to five disaster areas. According to the total disaster area, three warehouses from 20 warehouses should be selected as emergency supplies storage warehouses to provide emergency supplies to the disaster area. The initial values of the evaluation indicators of the 20 alternative points are shown in table 1.

Table 1. Initial value of evaluation index for alternative warehouse point

Alternative warehouse	Safety protection	Traffic convenience	Communication smoothness	Easy extensibility	geological conditions
1	0.56	0.79	0.89	0.82	0.68
2	0.89	0.76	0.65	0.79	0.68
3	0.75	0.76	0.45	0.79	0.89
4	0.56	0.89	0.79	0.59	0.75
5	0.83	0.89	0.85	0.89	0.89
6	0.89	0.75	0.82	0.56	0.46
7	0.75	0.89	0.89	0.75	0.46
8	0.56	0.68	0.59	0.75	0.75
9	0.87	0.89	0.85	0.86	0.86
10	0.83	0.86	0.89	0.82	0.85
11	0.46	0.79	0.45	0.82	0.75
12	0.89	0.82	0.84	0.82	0.86
13	0.86	0.86	0.89	0.86	0.89
14	0.68	0.56	0.79	0.46	0.75
15	0.87	0.88	0.82	0.85	0.89
16	0.89	0.89	0.85	0.89	0.86
17	0.25	0.45	0.68	0.89	0.89
18	0.58	0.59	0.76	0.79	0.86
19	0.76	0.39	0.76	0.56	0.46
20	0.92	0.59	0.25	0.68	0.59

According to the analysis and decision-making of experts, the weight values of five evaluation indexes are calculated by AHP as follows: 0.1707,0.2230,0.1781,0.2177,0.2105。Using the TOPSIS method based on B-type Association in literature [12], the closeness of 20 alternative emergency material storage warehouses is calculated as follows:

u_j=(0.32,0.41,0.32,0.28,0.66,0.43,0.43,0.42,0.49,0.49,0.44,0.63,0.64,0.21,0.58,0.79,0.47,0.58,0.65,0.34,0.48,0.28,0.41)。

The screening standard is set to be no less than 0.4, then the initial alternative warehouse for emergency materials reserve is: alternative warehouse 5, alternative warehouse 9, alternative warehouse 10, alternative warehouse 12, alternative warehouse 13, alternative warehouse 15 and alternative warehouse 16. According to GIS, the location information of 7 initial alternative emergency material storage warehouses, disaster areas and rear material storage warehouses can be obtained, as shown in table 2.

Table 2. Parameters of spare parts supply time for initial alternative field depot

Alternative warehouse	Disaster area					Rear material warehouse	Inventory capacity S_j
	1	2	3	4	5		
5	3	2.5	3	3	2.5	8	40
9	4	5.5	4	2.5	2	20	45
10	2	2.5	3	5.5	1.5	10	50
12	5	4	3.5	2	6	18	40
13	2.5	3.5	4	1.5	4	8	50
15	6.5	3.5	2	2.5	2	12	50
16	1.5	1.5	3.5	2.5	3.5	10	40
Emergency material demand	25	25	15	30	20	—	—
Maximum delay time	4	3.5	3.5	3	3	—	—
Importance degree	0.18	0.25	0.1	0.15	0.32	—	—
Penalty cost per unit time c_i	6	8	3	5	10	—	—
parameter α_i	0.6	0.75	0.3	0.5	0.9	—	—
parameter $\beta_{i'}$	1.2	1.3	1.05	1.1	1.5	—	—
parameter γ	1	1	1	1	1	—	—

According to formula (9) ~ (13), using the genetic algorithm simulation calculation, the optimal scheme of the maximum generalized coverage location model is: warehouse 10, warehouse 13, warehouse 16.

5. Conclusion

In this paper, considering the timeliness constraints and warehouse capacity constraints of emergency supplies in each disaster area, the satisfaction function of emergency supplies service and the penalty cost function of emergency supplies supply are constructed, and a multi-objective generalized maximum coverage location model is established. The optimal site selection scheme is calculated. The results show that the method proposed in this paper can solve the location problem of emergency material storage warehouse under the condition of multiple constraints in disaster area.

Acknowledgments

This work was financially supported by National natural science foundation of China (No.71871219) and National social science foundation of China (No.16gj003-069).

References

[1] LI J Y, WANG D, LIU Z, DAI X. (2018) A New Approach of Location Method for Logistics Park Based on Mixed Multi-Attribute Decision in Low Carbon Perspective. Journal of Shenyang University(Natural Science), 30:40-45.

[2] LI W J, CHEN J F, GUO C H, SHI Y, FAN X, LIU J F. (2018) Site Selection of Electricity Office in Rural Electric Power Business Area Based on AP Clustering and Set Covering Model. Mathematics in Practice and Theory, 48: 102-110.

[3] Sudtachat K, Mayorga M E, Mclay L A. (2016) A nested-compliance table policy for emergency medical service systems under relocation. Omega, 58: 154-168.

[4] LU Y Y, GUO Z J, WANG B L, ZHANG Q R, LU Q. (2018) A Traffic Rescue Centre Median Location Model Approaching Ideal Grey Relational Constraints. Journal of Highway and Transportation Research and Development, 35:118-123.

[5] ZHU X J, FENG X S, ZHENG Y, JIE Y P, HUA W X. (2019) Research on Location Optimization of Emergency Service Facilities. Journal of DALIAN jiaotong University, 40: 10-15.

[6] CHEN Z F, LI J W, LU F X, LI Q. (2018) Optimizing location of fire stations and it' s enlightenments for Xiong'an New Area. Journal of Safety Science and Technology, 14:14-19.
[7] CHEN J. (2016) Study on Support Decision of Tactical Ordnance Equipment Supply in Wartime. Army Engineering University, 32-36.
[8] SONG Y, TENG C M, JIANG J G. (2019) A Decision Model of Reserve Coverage Selection for Multi-level Service Facilities Based on Improved NSGA-II Algorithm. Operations Research and Management Science, 28: 75-82.
[9] YU D M, GAO L F, ZHAO S J. (2018) Location-allocation Optimization Model for Emergency Facilities and Algorithm with Maximum Time Satisfaction. Systems Engineering, 36: 99-106.
[10] LI M. (2010) The Research on A Method of The Tactical Maintenance Force Allocation in Wartime. Army Engineering University, 45-48.
[11] XIAO J H, HOU Y X. (2013) Research on emergency facility location model and algorithms for large-scale emergencies. Computer Engineering and Applications, 49: 67-71.
[12] SUN C J, PEI H, LIU D, LV W. (2018) Research on Post-earthquake Relief Cooperative Dispatching Considering Lateral Transshipment Policies. Journal of Wuhan University of Technology (Information & Management Engineering), 40: 389-395.

Optimization of Traffic Organization of loading ground based on transport loss

Mu Yang,

School of Traffic and Transportation, Beijing Jiaotong University, Beijing, 100044, China

Author's e-mail: 18140578@bjtu.edu.cn

Abstract. From the point of view of the railway company, the consignor and consignor to establish a cost optimization target system to study the loading area traffic organization optimization problem. Considering the transportation cost of the railway company and the cargo delivery side of the transport loss, this paper combines different characteristics of different cargo vulnerability and constructs a loading area traffic organization optimization model to minimize the total cost of the integrated economy.. Finally, a corresponding example is given, and the model is solved by using LINGO software.

1. Introduction

With the continuous expansion of the scale of the current railway network, railway transportation will be further developed. Considering the advantages of railway freight transportation, such as large volume, low cost, accurate arrival and departure time, railway transportation will be the first choice for the transportation of coal, oil, grain, aquatic products and other bulk goods. However, for different types of bulk goods, the transportation loss is different in the transportation process. We should combine the different loss characteristics of them to organize the relevant traffic as reasonably as possible, and establish a reasonable traffic organization scheme to ensure its reasonable and efficient transportation.

For the train flow organization, experts and scholars at home and abroad have done a lot of analyses and research, and achieved fruitful results. In the existing research fields, Literature [1] is the first study to use the systematic idea to put forward the "simultaneous calculation method" to optimize the traffic flow operation scheme of technical station. It established a model containing the information of train flow reorganization information in technical station, and the decision variable of the model is whether to organize direct freight trains or not. Literature [2] fully combines the necessary conditions and parameter selection for organizing the through train. It constructs a non-linear 0-1 programming model of the marshalling plan of the loading area, has a better description of the various combinations of the operation schemes of the starting train flow, and comprehensively analyzes the theoretical problems in many aspects of the marshalling plan of the loading area. Literature [3] and [4] consider both transportation cost and inventory cost, take the minimum cost of logistics system as the objective function at the same time, construct a non-linear 0-1 programming model for the direct freight train operation scheme in the loading area to determine the optimal scheme of train operation at the loading area. Literature [5] from the perspective of network flow and combination optimization, studies the comprehensive optimization problems of technical through (direct) marshalling plan, marshalling plan of through train in loading area of road network and train operation plan within district. Based on the OD traffic flow between the original stations, it deeply analyzes the advantages and disadvantages of operating section train, so as to build a comprehensive optimization model of direct to loading site, technical direct and district train formation plan. Finally, a numerical example is used to obtain the traffic organization optimization plan of minimum relative consumption of car-hour. Through comparative analysis, the rationality and necessity of bringing the operation conditions of district trains into the comprehensive optimization of train formation plan are verified. Literature [6] analyzes the categories, favorable indexes, operation conditions and scheme determination of the railway freight through train, and constructs the mathematical model of the marshalling plan of the direct transportation.

2. Problem description

The railway department generally takes the shortest operation car-hour as the optimization goal when preparing the train operation plan. The formulation of the train operation plan is considered from the perspective of the minimum transportation cost of the railway department. With the rapid development of society and economy, the competition among enterprises is becoming more and more fierce. In order to better adapt to the market demand, the railway department needs to optimize the plan from the overall interests of the consignee, consignor and railway department when preparing the train operation plan, so as to reduce the comprehensive cost of the whole logistics system to maximize the overall benefit.

The delivery of goods from the loading place to the unloading place must go through the delivery, transportation and receiving links. Each link will involve the respective interests of the consignor, the carrier and the consignee. Therefore, the target of the optimization of direct freight train schemes for loading area should be composed of the costs of the consignor, the carrier and the consignee. Among them, due to the different types of goods of the consignor and the consignee, the loss in the transportation is also different. For example, the loss of aquatic products, fruits and vegetables is far greater than the loss of coal and oil. Therefore, on the basis of the possibility and rationality of organizing the direct train at the loading place, the transportation loss of different goods shall be fully considered. With the goal of minimizing the total logistics cost, a 0-1 planning model of direct train operation scheme at the loading site is established. So as to provide a set of alternatives for the formulation of a reasonable organization scheme of traffic flow in the loading area, and finally realize the maximization of the interests of the railway company and customers.

3. Model construction

3.1. The Introduced Parameters are as follows:

N_{st} :average daily traffic volume from loading site's' to unloading site't'.

$\omega_{st_l}^{d}$:The consumption of car-hour of each vehicle of the initial departure traffic flow leaving the loading area in a direct way.

$\omega_{st_l}^{n-d}$:The consumption of car-hour of each vehicle of the initial departure traffic flow leaving the loading area in a non direct way.

$\omega_{st_{ul}}^{d}$:The consumption of car-hour of each vehicle of the initial departure traffic flow entering the unloading area in a direct way.

$\omega_{st_{ul}}^{n-d}$:The consumption of car-hour of each vehicle of the initial departure traffic flow entering the unloading area in a non direct way.

t_i:The average consumption of car-hour of each vehicle of the reorganization in station 'i' of the non direct train of the initial departure traffic flow deducting the average consumption of car-hour of each vehicle of direct train of the initial departure traffic flow without reorganization.

m_{st} :The maximum vehicle number of dense loading and unloading allowed in the initial departure traffic flow.

$\overline{m}_{st}$:The average number of vehicles in the initial departure traffic flow.

L_{st}:The economic loss of the goods owner caused by the delay of one hour for each vehicle in different initial departure traffic flow.

γ:Loss equivalent, the economic loss per hour of each vehicle converted into parking car-hour.

V:The collection of technical stations on the road network.

V(s):The collection of technical stations adjacent to the loading site.

Q(s):The collection of all unloading places of the traffic flow started from the loading place 's'.

V (k, t) :The collection of technical stations to be reorganized along the way, and it is agreed that 'k'station is included in V (k, t).

3.2. Decision Variables

1、 $x_{st} = \begin{cases} 1 & \text{If the traffic flow operates the direct train from loading site to unloading site} \\ 0 & \text{else} \end{cases}$

2、 $x_{st}^{k} = \begin{cases} 1 & \text{If the first reorganization station of Nst is 'k'} \\ 0 & \text{else} \end{cases}$

3、 $I_{st} = \begin{cases} 1 & \text{If } m_{st} \geq \overline{m}_{st} \\ 0 & \text{else} \end{cases}$

3.3. objective function

When the initial departure flow 'N_st' is a large train flow, the consumption of car-hour of the direct train from the loading place to the unloading place is:

$$Z_1 = N_{st}x_{st}(\omega_{st_l}^{d} + \omega_{st_{ul}}^{d}) \quad (1)$$

When the initial departure train flow 'Nst' is sent to the nearest technical station in front operated as the technical train flow, the converted consumption of car-hour generated at the loading place and unloading place is:

$$Z_2 = N_{st}x_{st}^{k}(\omega_{st_l}^{n-d} + \omega_{st_{ul}}^{n-d}), k \in V(s) \quad (2)$$

The initial departure train flow 'Nst' is sent to the adjacent technical station in front operated as the technical train flow. The consumption of car-hour of the train for operation in the technical station and the loss of goods during transportation:

$$Z_3 = N_{st}x_{st}^{k} \sum_{i \in V(k,t)} t_i (1 + L_{st}\gamma), k \in V(s) \quad (3)$$

When the initial departure train flow 'Nst' is sent to the non adjacent technical station in front operated as the technical train flow, the converted consumption of car-hour generated at the loading place and unloading place is as follows:

$$Z_4 = N_{st}x_{st}^{k}(\omega_{st_l}^{d} + \omega_{st_{ul}}^{n-d}), k \in V \backslash V(s) \quad (4)$$

When the initial departure train flow 'Nst' is sent to the non adjacent technical station in front operated as the technical train flow, the consumption of car-hour of the train for operation in the technical station and the loss of goods during transportation:

$$Z_5 = N_{st}x_{st}^{k} \sum_{i \in V(k,t)} t_i (1 + L_{st}\gamma), k \in V \backslash V(s) \quad (5)$$

Therefore, the total consumption of train flow on the way is:

$$Z = \sum_{i=1}^{5} Z_i \quad (6)$$

3.4. Constraints

There are three ways for the initial departure flow to be transported from the loading place to the unloading place:

- Organize the operation of direct trains from loading place to unloading place.
- Organize the operation of the direct train from the loading place to the non adjacent technical station in front, and merge into the corresponding technical flow after the technical station is reorganized.

- Organize the operation of the direct train from the loading place to the adjacent technical station in front, and merge into the corresponding technical flow after the technical station is reorganized.

Therefore, the uniqueness condition of the scheme of N_{st} organization of traffic flow is as follows.:

$$I_{st}x_{st} + \sum_{k \in V} x_{st}^{k} = 1 \tag{7}$$

I_{st} is to meet the necessary conditions of the traffic (large stocks of traffic) open line direct unloading of the train.

The allowable dense loading number m_{st} of the initial departure flow N_{st} is:

$$m_{st} = \min\{m_{st}^{L}, m_{st}^{U}\} \tag{8}$$

In the formula, m_{st}^{L} is the number of allowed dense loading of traffic flow N_{st}; m_{st}^{U} is the number of allowed dense unloading of traffic flow N_{st}.

For the large train flow, i.e. the train flow of $m_{st} \geq \overline{m}_{st}$ (here refers to the situation that the daily average planned train flow can meet the requirement of organizing the direct train independently), the direct train shall be organized separately. For the small train flow, i.e. the train flow with $m_{st} < \overline{m}_{st}$, it should be considered whether to combine with other train flows to operate the direct train. When a initial departure direct train operated from loading place s to a fulcrum station K, the total number of allowed dense loading and unloading vehicles of each flow attracted by the direct train destination shall meet the necessary conditions:

$$\overline{m}_{st} I\left(\sum_{t} x_{st}^{k}\right) - \sum_{t} m_{st} x_{st}^{k} \leq 0 \tag{9}$$

In the formula: $I(x)$ is a step function, defined as:

$$I(x) = \begin{cases} 1, & x > 0 \\ 0, & x \leq 0 \end{cases} \tag{10}$$

Because this is a nonlinear 0-1 programming model. In order to solve the model conveniently, it is linearized here. May order:

$$x_{sk} = I\left(\sum_{t} x_{st}^{k}\right) = \begin{cases} 1 & \text{there is a direct train from loading station s to technical station} \\ 0 & \text{others} \end{cases} \tag{11}$$

Bring the formula in, then we can get :

$$\overline{m}_{st} x_{sk} - \sum_{t} m_{st} x_{st}^{k} \leq 0 \tag{12}$$

As long as $x_{st}^{k} > 0$, , there must be $x_{sk} = 1$. Therefore, a set of constraints should be added accordingly after introducing the variable x_{sk}:

$$x_{sk} \geq x_{st}^{k} \tag{13}$$

Using the characteristics of 0-1 variables, this set of constraints is superposed, and the result is as follows:

$$Mx_{sk} - \sum_{t} x_{st}^{k} \geq 0 \tag{14}$$

In summary, the model can be converted to the following linear 0-1 planning form:

$$\text{Min} \quad Z = \sum_{i=1}^{5} Z_i$$

$$\text{s.t.} \begin{cases} I_{st}x_{st} + \sum\limits_{k \in V} x_{st}^{k} = 1 & \forall t \in Q(s) \\ \overline{m}_{st} I\left(\sum\limits_{t} x_{st}^{k}\right) - \sum\limits_{t} m_{st} x_{st}^{k} \leq 0 & k \in V \backslash V(s) \\ Mx_{sk} - \sum\limits_{t} x_{st}^{k} \geq 0 & k \in V \backslash V(s) \\ \sum\limits_{l} y_{ij}^{l} = 0 & \forall i, j \\ x_{st}, x_{st}^{k}, x_{sk} \in \{0,1\} & \end{cases}$$

IOP Conf. Series: Materials Science and Engineering **792** (2020) 012014 doi:10.1088/1757-899X/792/1/012014

4. Calculating examples

A simple road network structure diagram is given here, as shown in Figure 1 below, where 1 represents the loading station of the departure station, 2-9 represent the technical station on the way, and 10-14 represent the unloading station of the terminal station.

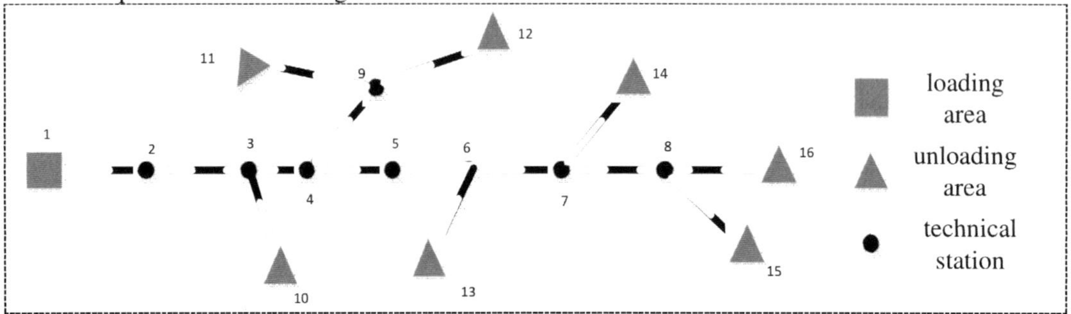

Figure 1 Network structure

4.1. The relevant data and parameters

Table 1 Traffic OD data

Car flow N_{st}	Cars number (car/day)	m_{st} (car)	Consumption γL_{st} (car per hour)
N_{10}	100	100	0.1
N_{11}	12	12	1.8
N_{12}	32	32	0.3
N_{13}	30	30	1.9
N_{14}	42	42	1.2
N_{15}	53	53	1.1
N_{16}	34	34	0.7

4.2. Technical parameters of each technical station

Table 2 Technical station parameter

Technical Station Number	2	3	4	5	6	7	8	9	2
Technical station reoperation consumption t_i	5	3	4	3	4	4	5	5	5
$\overline{m}_{sk}$	40	40	40	40	40	40	40	40	40

4.3. Other parameters

The cost of loading and unloading is as follows::

$$\omega_{st_l}^{d} = 10,\ \omega_{st_l}^{n-d} = 15,\ \omega_{st_{ul}}^{d} = 8,\ \omega_{st_{ul}}^{n-d} = 18.$$

4.4. Example model

To simplify the calculation, this example shows a linear road network structure with only one loading place. Therefore, only the unloading place is marked for the initial departure flow, and the underground loading mark is omitted. The example model is shown as follows:

$$Min = C_{10} + C_{11} + C_{12} + C_{13} + C_{14} + C_{15} + C_{16}$$

In the objective function, the expressions of C_{10}, C_{11}, C_{12}, C_{13}, C_{14}, C_{15}, C_{16} are respectively as follows:

$$C_{10}=N_{10}\left\{\left(w_{10l}{}^{d}+w_{10ul}{}^{d}\right)x_{10}+\left[w_{10l}{}^{nd}+w_{10ul}{}^{nd}+\sum_{i=V(2,10)}t_i(1+L_{10}\gamma)\right]x_{10}^{2}+\left[w_{10l}{}^{nd}+w_{10ul}{}^{nd}+\sum_{i=V(3,10)}t_i(1+L_{10}\gamma)\right]x_{10}^{3}\right\}$$

$$C_{11}=N_{11}\left\{\left(w_{11l}{}^{d}+w_{11ul}{}^{d}\right)x_{11}+\left[w_{11l}{}^{nd}+w_{11ul}{}^{nd}+\sum_{i=V(2,11)}t_i(1+L_{11}\gamma)\right]x_{11}^{2}+\sum_{k=3,4,9}\left[w_{11l}{}^{nd}+w_{11ul}{}^{nd}+\sum_{i=V(k,11)}t_i(1+L_{11}\gamma)\right]x_{11}^{k}\right\}$$

$$C_{12}=N_{12}\left\{\left(w_{12l}{}^{d}+w_{12ul}{}^{d}\right)x_{12}+\left[w_{12l}{}^{nd}+w_{12ul}{}^{nd}+\sum_{i=V(2,12)}t_i(1+L_{12}\gamma)\right]x_{12}^{2}+\sum_{k=3,4,9}\left[w_{12l}{}^{nd}+w_{12ul}{}^{nd}+\sum_{i=V(k,12)}t_i(1+L_{12}\gamma)\right]x_{12}^{k}\right\}$$

$$C_{13}=N_{13}\left\{\left(w_{13l}{}^{d}+w_{13ul}{}^{d}\right)x_{13}+\left[w_{13l}{}^{nd}+w_{13ul}{}^{nd}+\sum_{i=V(2,13)}t_i(1+L_{13}\gamma)\right]x_{13}^{2}+\sum_{k=3}^{6}\left[w_{13l}{}^{nd}+w_{13ul}{}^{nd}+\sum_{i=V(k,13)}t_i(1+L_{13}\gamma)\right]x_{13}^{k}\right\}$$

$$C_{14}=N_{14}\left\{\left(w_{14l}{}^{d}+w_{14ul}{}^{d}\right)x_{14}+\left[w_{14l}{}^{nd}+w_{14ul}{}^{nd}+\sum_{i=V(2,14)}t_i(1+L_{14}\gamma)\right]x_{14}^{2}+\sum_{k=3}^{7}\left[w_{14l}{}^{nd}+w_{14ul}{}^{nd}+\sum_{i=V(k,14)}t_i(1+L_{14}\gamma)\right]x_{14}^{k}\right\}$$

$$C_{15}=N_{15}\left\{\left(w_{15l}{}^{d}+w_{15ul}{}^{d}\right)x_{15}+\left[w_{15l}{}^{nd}+w_{15ul}{}^{nd}+\sum_{i=V(2,15)}t_i(1+L_{15}\gamma)\right]x_{15}^{2}+\sum_{k=3}^{8}\left[w_{15l}{}^{nd}+w_{15ul}{}^{nd}+\sum_{i=V(k,15)}t_i(1+L_{15}\gamma)\right]x_{15}^{k}\right\}$$

$$C_{16}=N_{16}\left\{\left(w_{16l}{}^{d}+w_{16ul}{}^{d}\right)x_{16}+\left[w_{16l}{}^{nd}+w_{16ul}{}^{nd}+\sum_{i=V(2,16)}t_i(1+L_{16}\gamma)\right]x_{16}^{2}+\sum_{k=3}^{8}\left[w_{16l}{}^{nd}+w_{16ul}{}^{nd}+\sum_{i=V(k,16)}t_i(1+L_{16}\gamma)\right]x_{16}^{k}\right\}$$

S.T

$$I_{10}x_{10}+x_{10}^{2}+x_{10}^{3}=1$$

$$I_{11}x_{11}+x_{11}^{2}+x_{11}^{3}+x_{11}^{4}=1$$

$$I_{12}x_{12}+x_{12}^{2}+x_{12}^{3}+x_{12}^{4}+x_{12}^{5}=1$$

$$I_{13}x_{13}+x_{13}^{2}+x_{13}^{3}+x_{13}^{4}+x_{13}^{5}+x_{13}^{6}=1$$

$$I_{14}x_{14}+x_{14}^{2}+x_{14}^{3}+x_{14}^{4}+x_{14}^{5}+x_{14}^{6}+x_{14}^{7}=1$$

$$I_{15}x_{15}+x_{15}^{2}+x_{15}^{3}+x_{15}^{4}+x_{15}^{5}+x_{15}^{6}+x_{15}^{7}+x_{15}^{8}=1$$

$$I_{16}x_{16}+x_{16}^{2}+x_{16}^{3}+x_{16}^{4}+x_{16}^{5}+x_{16}^{6}+x_{16}^{7}+x_{16}^{8}=1$$

$$\overline{m_3}x_3-m_{10}x_{10}^{3}-m_{11}x_{11}^{3}-m_{12}x_{12}^{3}-m_{13}x_{13}^{3}-m_{14}x_{14}^{3}-m_{15}x_{15}^{3}-m_{16}x_{16}^{3}\le 0$$

$$\overline{m_4}x_4-m_{11}x_{11}^{4}-m_{12}x_{12}^{4}-m_{13}x_{13}^{4}-m_{14}x_{14}^{4}-m_{15}x_{15}^{4}-m_{16}x_{16}^{4}\le 0$$

$$\overline{m_5}x_5-m_{13}x_{13}^{5}-m_{14}x_{14}^{5}-m_{15}x_{15}^{5}-m_{16}x_{16}^{5}\le 0$$

$$\overline{m_6}x_6-m_{13}x_{13}^{6}-m_{14}x_{14}^{6}-m_{15}x_{15}^{6}-m_{16}x_{16}^{6}\le 0$$

$$\overline{m_7}x_7-m_{14}x_{14}^{7}-m_{15}x_{15}^{7}-m_{16}x_{16}^{7}\le 0$$

$$\overline{m_8}x_8-m_{15}x_{15}^{8}-m_{16}x_{16}^{8}\le 0$$

$$\overline{m_9}x_9-m_{11}x_{11}^{9}-m_{12}x_{12}^{9}\le 0$$

$$7x_3-x_{10}^{3}-x_{11}^{3}-x_{12}^{3}-x_{13}^{3}-x_{14}^{3}-x_{15}^{3}-x_{16}^{3}\ge 0$$

$$6x_4-x_{11}^{4}-x_{12}^{4}-x_{13}^{4}-x_{14}^{4}-x_{15}^{4}-x_{16}^{4}\ge 0$$

$$4x_5-x_{13}^{5}-x_{14}^{5}-x_{15}^{5}-x_{16}^{5}\ge 0$$

$$4x_6-x_{13}^{6}-x_{14}^{6}-x_{15}^{6}-x_{16}^{6}\ge 0$$

$$3x_7-x_{14}^{7}-x_{15}^{7}-x_{16}^{7}\ge 0$$

$$2x_8-x_{15}^{8}-x_{16}^{8}\ge 0$$

$$2x_9-x_{11}^{9}-x_{12}^{9}\ge 0$$

$$\forall x_{st}, x_{st}^{k}, x_{sk} \in \{0,1\}$$

$$\forall x_{st}, x_{st}^{k}, x_{sk} \in \{0,1\}$$

4.5. Using LINGO commercial software to solve the above model, the results are as follows:

The results obtained are as follows: $x_{10}=1$, $x_{14}=1$, $x_{15}=1$, $x_{11}^{9}=1$, $x_{12}^{9}=1$, $x_{13}^{6}=1$, $x_{16}^{6}=1$ The other variables are all 0, and the cost of this scheme is 6306 car-hour.The corresponding traffic organization scheme is shown in Figure 2 below:

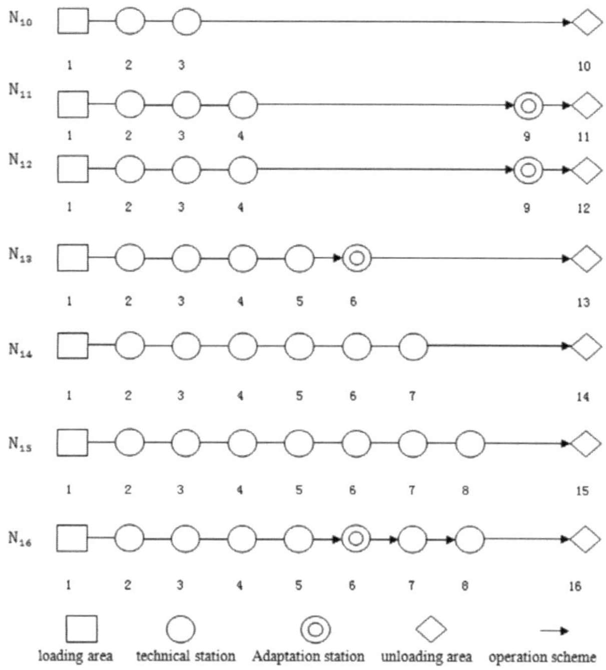

Figure 2 Traffic Organization programme

In the figure above, the trains that arrive at 10, 14 and 15 unloading stations and arrive at other unloading stations will be reorganized and delivered. Among them, the trains that arrive at 11 and 12 unloading stations will be reorganized at 9 technical station on the way, and the trains that arrive at 13 and 16 unloading stations will be reorganized at 6 technical station on the way.

5. Summary

For the transportation of railway bulk goods, different forms of train flow organization will produce different transportation and loading costs. Different types of goods can bring different transportation loss costs. Based on the point-to-point transportation organization, this paper analyzes the conditions of the point-to-point direct trains, establishes the point-to-point optimization organization model, selects the point-to-point direct trains or other forms of trains according to their different restrictions, and finally gives the relevant examples for analysis, thus verifying the feasibility of the model.

References

[1] H.E polovoi. Transportation of goods to[M]. Trans. Teaching and Research Office of traffic organization, Northern Jiaotong University.Beijing: China Railway press, 1981.
[2] Lin Boliang, Chu Songnian, Shi DeYao, He Shiwei. Optimization model of marshalling plan of through train in loading area[J].China Railway science, 1995 (02):108-114.
[3] Cao X, Lin B, Liu H, Yan H.Optimization of direct traffic organization of the base[J].Railway Journal, 2007 (01):16-20.

[4] Cao X, Lin B, Yan H.Optimization model of direct train operation scheme in loading area [J].Railway Journal, 2006 (04):6-11.
[5] Wang L, Ma J, Lin B, Ni S, Lv H.Comprehensive optimization of direct and technical train formation plan in loading area and Section [J].Transportation system engineering and information,2013,13(05):127-133.
[6] Fan Z, Lin B, Niu H.Study on the plan of starting from loading place to direct operation [J].Logistics technology, 2007 (06):77-80.

IOP Conf. Series: Materials Science and Engineering **792** (2020) 012015 doi:10.1088/1757-899X/792/1/012015

Study on the Distribution Law and Speed Limit of Night Speed in Expressway Maintenance Area

Yong Huang[1*], Daqing Fang[2], Fuliang Tan[2], Minglu Tao[2], Daoguang Si[2], Yang Shu[2]

[1,2]Anhui Transportation holding Group Co., Ltd, Hefei, Anhui, 230001, China

[*]Corresponding author's e-mail: 3055768945@qq.com

Abstract. In order to reasonably determine the maximum and minimum vehicle speed limit values at night and ensure the driving safety in the maintenance area of expressway, the distribution law of vehicle speed at night in the maintenance area of expressway is studied .Based on the investigation of one-way lane closure construction in the four-lane maintenance area of expressway, the location speed of the main traffic control area in the night maintenance area is analyzed, and the average speed, 85% vehicle speed, the speed difference in different areas and the speed difference in heavy-duty vehicle(HDV) and light-duty vehicle(LDV) workshops are determined. The results show that the standard deviation of vehicle speed distribution from warning area to transition zone to construction area is obviously reduced, and the speed difference between HDV and LDV is also obviously reduced. Finally, according to the distribution characteristics of night speed, the night speed limit model of maintenance operation area is given, the speed limit values of different models are calculated respectively, and the night perceived speed model is used to modify the speed limit value.

1. Introduction

Because of the closed lane, the traffic capacity of the expressway vehicle maintenance area will be reduced, which is easy to cause traffic accidents. Especially when driving at night in the maintenance area of expressway, the driver's vision will be reduced by about 50% [1]. According to statistics, the traffic accidents in the maintenance area at night account for about 60% of the total traffic accidents in the maintenance area. Therefore, it is of great significance to study the speed distribution and speed limit in the day and night expressway maintenance area.

At present, the existing literature has also studied the speed distribution law of each area in the expressway maintenance area.Bella [2] collected the driving simulator to study the night speed distribution. Meng [3] has made clear the speed distribution law of different types of maintenance areas of expressway through statistical analysis of measured speed data. At the same time, for the research on the speed limit of highway maintenance operation area, many research results have been obtained abroad, mainly related to the speed change law, variable speed limit control and real-time speed limit control in various areas of the operation area [4]. Domestic scholars mainly study the static speed limit problem and its evaluation of expressway maintenance area [5].

To sum up, there is a lack of research on speed distribution and speed limit at night in expressway maintenance area. Therefore, based on the measured speed detection data of one-way lane closed maintenance area of Anhui two-way four lane expressway, this paper determines the speed distribution law of each section of one-way lane closed construction of day and night expressway, and gives a reasonable speed limit value. At the same time, this paper uses the factors that affect the capacity of expressway maintenance area to modify the speed limit value.

2. Field investigation of test section

In order to study the law of speed distribution at night in each area of expressway maintenance work area, the field traffic investigation of pavement maintenance of Shanghai-Shanxi Expressway (G40) was carried out. The expressway maintenance area is located in K120-122 section of Shanghai Shaanxi Expressway.The traffic volume and vehicle type data of the test section are collected by the automatic traffic volume survey instrument, and the survey location is the upstream warning area, transition area and work area. The location of the investigation is shown in Figure 1. At the same time, in the maintenance area where there are relatively few vehicles at night, from 22:00 to 24:00, the driving speed of HDV and LDV are recorded by driving recorder, and the recognition distance and perceived speed of different vehicles at night are analyzed.

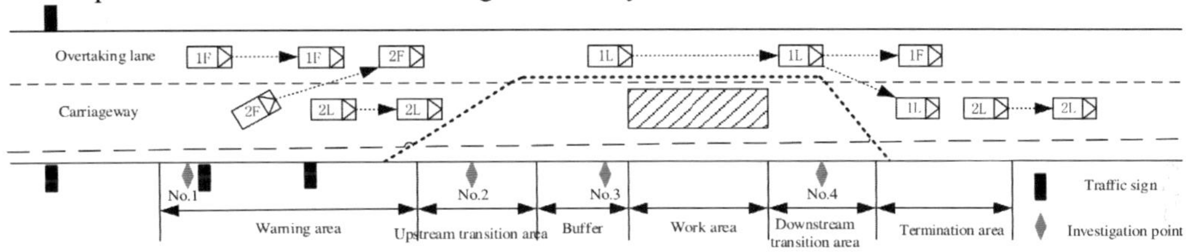

Figure 1. Investigation site for closed maintenance operation area of highway single-lane road

Combined with the technical standard for Highway Engineering (JTG B01-2014) and the vehicle type division method of the automatic traffic volume survey instrument, motorcycle and minibus belong to LDV; other vehicles are HDV. MATLAB is used to process the night speed of the investigation points in each traffic control area of the investigation section, and the analysis results are shown in Table 1.

Table 1. Statistical parameters for the vehicle speed at the site of investigation

Traffic control area	Vehicle type	Sample size	Statistical parameters of night speed (km/h)						
			V15	V50	V85	Vmin	Vmax	Average	Standard deviation
warning area	HDV	865	28.6	74.1	91.3	10.6	145.4	75.6	32.9
	LDV	2136	34.1	75.0	95.1	10.8	148.6	79.7	34.6
Transiti on area	HDV	538	15.7	49.6	82.7	9.0	112.7	52.3	29.4
	LDV	2496	18.3	47.2	83.9	9.0	130.1	53.1	28.1
Work area	HDV	476	54.6	66.4	77.1	25.5	85.2	66.1	10.58
	LDV	2378	52.6	64.6	78.6	15.8	102.6	65.3	11.24

In order to show the short-term fluctuation of traffic flow, this paper sets the time interval as 5min to calculate the average value, the 15th% speed, the 50th% speed, the 85% speed and the standard deviation of vehicle speed of each group of samples.

3. Speed distribution law of each area in the night maintenance operation area

This paper takes the speed as the horizontal axis and the probability density as the vertical axis to fit the speed distribution at night, and draws the speed probability density curve of warning area, transition area, small and large vehicles in the maintenance area, as shown in Figure 2.

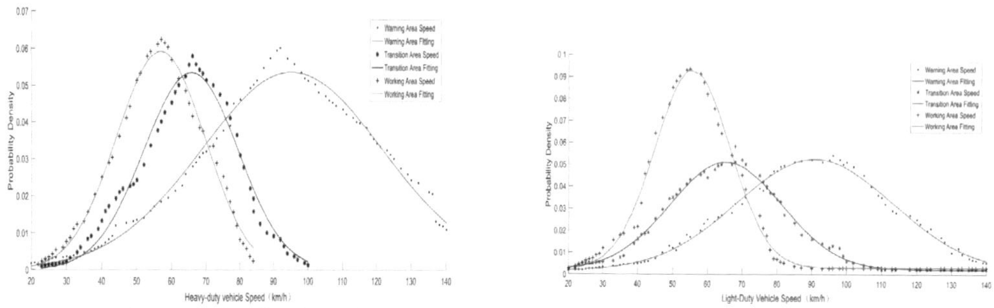

(a) Normal speed distribution of HDV (b) Normal speed distribution of LDV

Figure 2. Normal distribution of vehicle speed in different traffic control areas of maintenance operation area

It can be seen from Table 2 that when the vehicles pass through the warning area, upstream transition area and work area in turn, the 85% speed of small vehicles gradually decreases, and small vehicles are speeding in each area. The 85% speed of large vehicles also shows a downward trend, and HDV are also exceeding the speed limit in each area.

Table 2. Traffic control area running speed statistics

Vehicle type	Traffic control area		
	Warning area	Upstream transition	Work area
LDV	91.3	82.7	77.1
HDV	95.1	82.9	78.6

4. Research on recognition distance and perceived speed of maintenance area based on speed distribution law

4.1. Night recognition distance model of different models

According to the speed distribution law in the maintenance area, the night recognition distance of different vehicle models at different speeds is obtained through experiments, as shown in Table 3. According to Table 3, the maximum recognition distance of HDV and LDV is 305m and 251m, and the minimum values are 243m and 188m. And the recognition distance of large vehicle is far from that of small vehicle. The recognition distance of night maintenance work area decreases with the increase of vehicle speed.

Table 3. Night recognition distance of different vehicle types

Vehicle type	Night speed（km/h）						
	40	50	60	70	80	90	100
HDV	183	175	167	158	146	134	128
LDV	181	173	162	148	140	130	120

Through the fitting analysis of the test data, the change rule of night recognition distance and speed of different models is shown in Figure 4. According to Figure 3, with the increase of vehicle speed, the recognition distance at night decreases gradually, and the recognition distance is gradually in the middle. By linear regression analysis of recognition distance with MATLAB software, the function relationship and correlation coefficient R of different vehicle types are 0.9939 and 0.9908 respectively. The expression of night recognition distance model of different models is as follows:

$$D(v) = -0.9571v + 222.9 \quad (1)$$

$$d(v) = -1.039v + 223.3 \quad (2)$$

Where: $D(v)$ is the night recognition distance of HDV; $d(v)$ is the night recognition distance of LDV.

IOP Conf. Series: Materials Science and Engineering **792** (2020) 012015 doi:10.1088/1757-899X/792/1/012015

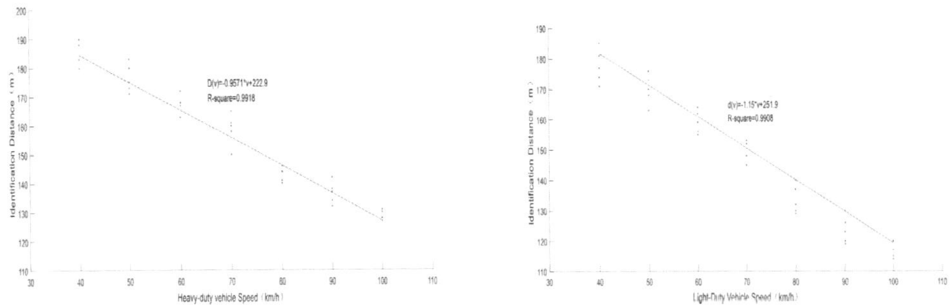

(a) One term fitting of HDV (b) One term fitting of LDV

Figure 3. Night recognition distance and speed fitting of different vehicle types

4.2. Night sensing speed model of different models

The paper analyzes the perceptual speeds of different models at different vehicle speeds at night, as shown in Table 4. It can be concluded from Table 4 that when the speed of HDV and LDV in the maintenance operation area is greater than 80km/h, the night-time perceived speed of different models is basically the same as the vehicle speed. According to figure 4, perceived speed is directly proportional to the actual speed. Regression analysis with MATLAB software shows that the correlation coefficients R of different models are 0.9871 and 0.9837, respectively. The results show that perceived speed is significantly related to the actual speed. The expression of night perceived speed model of different models is as follows:

$$F(v) = 1.161v - 15.54 \tag{3}$$

$$f(v) = 1.232v - 21.96 \tag{4}$$

Where: $F(v)$ is the night perception speed of HDV; $f(v)$ is the night perception speed of LDV.

Table 4. Night perception speed of different vehicle types

Vehicle type	Actual speed at night（km/h）						
	40	50	60	70	80	90	100
HDV	35	40	50	65	80	90	100
LDV	30	40	45	65	80	90	100

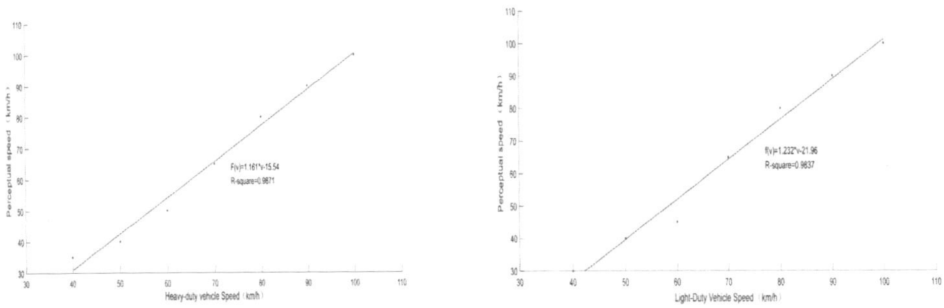

(a) Perception speed fitting of HDV (b) Perception speed fitting of LDV

Figure 4. The night perceived speed of HDV and LDV matches the actual speed

5. Study on speed limit at night in maintenance area

5.1. Speed limit model at night

In order to ensure safe driving, the vehicle identification distance must be greater than or equal to the stopping sight distance:

$$S_D \geq S_t + S_0 \tag{5}$$

Where: S_D is the identification distance; S_t is the stopping sight distance; S_0 is the safety distance（S_0=3m）.

Stopping sight distance:

$$S_t = \frac{v}{3.6}T + \frac{v}{7.2}t + \frac{v^2}{25.92a_{max}} \tag{6}$$

Where: T is the braking reaction time (T = 2.5s); t is the braking force rise time (t = 0.2s), a_{max} is the maximum braking deceleration （$a_{max} = g(\varphi + i)$）; φ is the road adhesion coefficient; i is the longitudinal slope.

Substituting formula (1), (2) into formula (5), the night speed limit models of large vehicles (7) and small vehicles (8) can be obtained:

$$v_D = [-0.84 + \sqrt{0.7 + 8.48/g(\varphi+i)}] \times 25.92g(\varphi+i) \tag{7}$$

$$v_d = [-0.88 + \sqrt{0.77 + 8.49/g(\varphi+i)}] \times 25.92g(\varphi+i) \tag{8}$$

Where: v_D , v_d are the maximum speed limit values at night in the maintenance area of HDV and LDV.

5.2. *Night speed limit correction based on perceived speed*

In conclusion, due to the influence of speed, road adhesion coefficient, gradient, etc., vehicle speed perception will be deviated during driving, so the speed limit value will be corrected. According to formulas (3) and (4), the large and small vehicle speed deviations (9) and (10) can be obtained as follows:

$$\Delta v_D = v_D - (1.16v - 15.54) \tag{9}$$

$$\Delta v_d = v_d - (1.232v - 21.96) \tag{10}$$

Where: Δv_D , Δv_d is the speed deviation of HDV and LDV.

Taking the speed deviation as the correction value of the night maintenance operation area, the maximum speed limit value of the night operation area after correction can be obtained, namely:

$$v_{mD} = v_D + \Delta v_D \tag{11}$$

$$v_{md} = v_d + \Delta v_d \tag{12}$$

Where: v_{mD}, v_{md} is the maximum speed limit correction value of HDV and LDV in the maintenance area.

6. Case analysis

In this paper, the longitudinal slope is set to 0, different pavement adhesion coefficient is selected, and MATLAB software is used to solve the model, as shown in Table 6 (the result is an integer of 5). It can be seen from Table 5 that: under the same model, the speed limit value decreases with the decrease of the adhesion coefficient; when the adhesion coefficient is the same, the speed limit value of LDV is less than that of HDV, which may be caused by the distance between HDV and LDV at night. In addition, the calculated maximum speed limit is 95km/h (mainly due to the small traffic flow in the maintenance operation area), which is consistent with the distribution of the maximum running speed of vehicles at 100km/h. Therefore, the speed limit model is basically consistent with the actual traffic conditions.

Table 5. Calculation of speed limit at night maintenance work area

Vehicle type	Speed limit recommendation（km/h）				
	0.6	0.5	0.4	0.25	0.15
HDV	95	90	85	75	65
LDV	90	85	80	70	60

7. Conclusions

The night speed limit model of the working area is established by the night recognition distance and the perceived speed model. The maximum speed limit value of the work area is 95km/h and the minimum speed limit value is 60km/h, which is basically the same as the speed distribution of the vehicle in the maintenance area. Based on the analysis of measured data, this paper determines the distribution law of vehicle speed in night maintenance operation area and gives the method of speed limit value and correction.

Acknowledgments

The project relies on the technology project of Anhui Transportation holding Group Co., Ltd.(Project No:GSKY-2019-01).

References

[1] Lin J Min.(2002)Visual characteristics of drivers and driving at night[J].Automobile and safety, 3:66-69.

[2] Bella F , Calvi A , D"Amico F . (2014)Analysis of driver speeds under night driving conditions using a driving simulator[J]. Journal of Safety Research, 49:45.e1-52.

[3] Meng X H, Shi Y Y,Wang H, et al. (2013)Speed distribution characteristics and speed limits of freeway work zones[J]. Journal of Transportation Systems Engineering and Information Technology, 13(1):153-159.

[4] Ravani B , Wang C . (2018)Speeding in highway work zone: An Evaluation of methods of speed control.[J]. Accid Anal Prev, 113:202-212.

[5] Wang Q, Wang X P. (2010)Research on speed limit control in expressway maintenance construction area[J].Journal of Transport Information and Safety,28(1) : 124-129.

Research on planning and design of airport airfield area

Li Ke*, **Shao Bin**

Aviation Engineering Institute, Air Force Engineering University, Xi'an 710038, China

*Corresponding author's e-mail: lkkgd12@163.com

Abstract: Airport status in city construction, regional layout, and transportation systems is analyzed. Influence of operation and management in military-civilian airports is pointed out from the perspective of airport planning and design and operation management. The airfield planning and design are analyzed from the capacity of runway, taxiway, and apron. The displaced threshold can meet the needs of the aviation business, while taking reconstruction and expansion of airport.

1. Introduction

Airport is a complex system engineering and a part of city[1]. Airport planning is an important part of city planning. As a sub-unit of the regional airport group, the airport must be integrated into the overall layout of the national airport construction and closely related to the management and optimization of the airspace. Airport is a comprehensive transportation system and the hub of transportation mode conversion, undertaking the transportation of passengers and cargo. The airport mainly includes airfield area, passenger terminal area, cargo terminal area, complex supporting area, aircraft maintenance area, and other functional areas[2]. The purpose of airport planning and design is to ensure the smooth operation of the airport. The operation and management mode of the airport affects the planning and design of the airport, and the planning and design of the airport and operation management should be combined as a whole.

2. Airport planning and design based on operation and management

The most important function areas of the airport are the airfield area, passenger terminal area and cargo terminal area, which are the core parts of airport. Airport planning is to coordinate the relationship between these functional areas. The operation and management mode of airport directly determines the planning and design of airport facilities. While planning and designing the airport, we must consider the operation and management mode of the airport.This is Liu Wu-jun's summary on the relationship between planning, design and operation, management of civil transport airport in "airport planning"[3]. Compared with civil airports, military airports or military-civilian airports have relatively few flight sorties and have certain particularity, but the airport planning and design must be studied from the perspective of operation management. The following analysis is carried out by taking the military-civilian airport as an example.

First, the effect of wind on takeoff and landing. According to relevant regulations and standards, the aircraft should not land on the side of the wind or take off and land with the wind. It is better to take off and land against the wind, but the headwind should not be too strong. When determining the direction of the runway, the guarantee rate shall not be less than 95%[1]. When military aircraft flight training and civilian aircraft flight operation are organized in the military-civilian airports, commanders and controllers are usually located in the military tower for flight command and aviation control. When spring meets summer and autumn meets winter, the wind direction is often changeable. During the flight training of military aircraft, if the wind direction changes, the command tower usually needs to be replaced to realize the requirements of taking off and landing against the wind during the flight interval. Compared with civilian aircrafts, military aircrafts are lighter in mass and more sensitive to wind. Civilian aircrafts may not need to change the direction of takeoff and landing, while military aircrafts have to change the direction of takeoff and landing. In the main landing of the runway, the communication and navigation facilities are more accurate and perfecter, the visual aids are higher, and the airport clearance conditions are well. However, at the secondary landing of the runway, the support facilities are relatively poorer. The temporary change of the command tower increases the risk of flight, to some extent.

Second, while military aircrafts usually take off continuously in batches at a short time, civil aircrafts usually leave or arrive at the airports according to the scheduled time of flight,and have to wait for takeoff or landing. In this case, it is not only necessary to adjust the departure time and approach landing time of the aircraft from the perspective of air traffic control, but also to analyze the taxi route of the aircraft before departure and after arrival, and reasonably plan the layout and configuration of the runway and taxiway of the airfield area. At a military-civilian airport, the crossings of taxi routes between military aircrafts and civil aircrafts reduce the operating efficiency of the airfield and increase the traffic risk on the ground. This situation is shown in Figure 1.

Third, from the viewpoint of common management of military and civil, the airfield area layout is optimized. For example, a parallel taxiway for civilian and military use is set up in front of the military apron. When a civilian aircraft turns to enter the vertical taxiway, its tail jet stream will affect the aircraft parked on the military apron and the staff for maintenance support. At a military-civilian airport, the situation that the road used by military command vehicles and the taxiway used by civilian aircraft are crossing is appeared , which brings great security risks for the operation and management of the airfield area, so airport authority has to send security personnel 24 hours on duty. An airfield layout is shown in Figure 1. From the perspective of planning and safety, it is necessary to make overall planning, comprehensive consideration and multi-use of all kinds of airfield roads, such as parallel roads used by troops, service lanes used by ground vehicles, and common patrol roads, to reduce the crossing of roads and taxiways, to avoid security threats, and to cut investment appropriately.

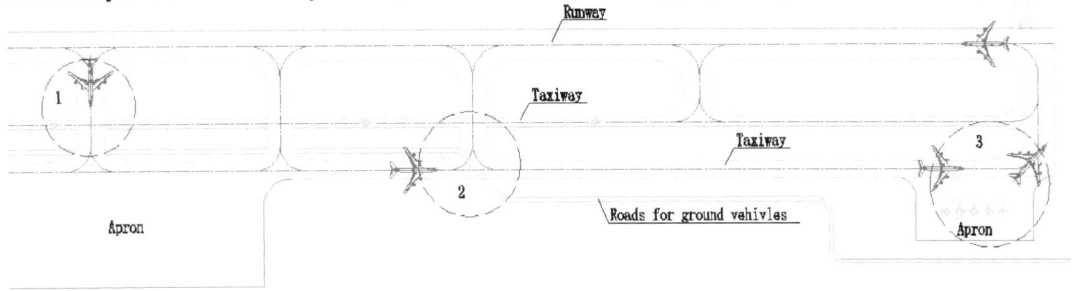

Figure 1. An example of airfield layout

3. Airfield area planning and design based on airport capacity

Airfield area planning and design is the most important content of airport planning and design, which is directly related to aircrafts and vehicle activities, operation management, and command and support. The capacity of an airport shall take into account the capacity of the airfield area, passenger terminal area, cargo terminal area, and passenger collection and distribution system. Capacity is based on an analysis of demand to meet the needs of the airline business. An important goal of airport capacity analysis is to make the capacity of each system without bottleneck.

Airfield capacity is one of the key factors restricting airport capacity. Airfield capacity is the capacity of serving aircraft sorties of airport facilities, which mainly manifest the capacity of runway, taxiway, and apron. Factors which affect the capacity of the airfield, include airfield geometry, usable airspace, air traffic control procedures, combination of aircraft types, weather conditions, operational command

and support capabilities[4]. The configuration and amount of runways are the key to determine the capacity of airfield. The capacity of the airfield should be built according to the near and long term runway capacity and the demand of gates, according to the principle of "unified planning and phased construction"[4][5].

3.1. Runway capacity

The number of runways almost doubles the capacity of runways, and the corresponding taxiway system is also needed. For example,one runway and one taxiway can handle 10 million passengers a year. When passenger throughput reaches 20 million, one runway is strained and two parallel taxiways are needed. More than 30 million people will need to build two runways. More than 40 million people need three runways[3]. The runway capacity of general airports is shown in Table 1.

Table 1. Runway capacity of general airports[3]

Serial number	Facilities	Capacity (person-times/year)
1	1-R/W、1-A/T/W	1000,000
2	1-R/W、1-T/W、1-A/T/W	2000,000
3	2-R/W (Closely-spaced Parallel Runway)、2/3-T/W、1-A/T/W	3000,000-4000,000
4	2-R/W、2-T/W、1/2-A/T/W	4000,000
5	2-R/W、4-T/W、? -A/T/W	4000,000-6000,000
6	3-R/W(Closely-spaced Parallel Runway)、3-T/W、2-A/T/W	5000,000-6000,000

Note: 1. The above is the ultimate capacity of the runway, without considering the influence of other factors;
2. A/T/W is is short for Apron Taxi Way;
3. ? indicates that the number is uncertain.

Runway spacing also has a significant impact on runway capacity. The two runways are too close to be used independently. One must land and the other must take off. If the distance between the two runways is increased to 1035 meters, the two runways can operate independently[2].

3.2. Capacity of taxiway system

In taxiway system, the rapid exit taxiway has a great influence on the increase of runway capacity. The rapid exit taxiway is usually at an Angle of 30 degrees away from the runway. The aircraft can leave the runway at a higher speed because the second aircraft can only use the runway after the previous one has left the runway.

In a complex multi-runway airport, taxiways cross the runway, affecting the take-off or landing of other aircraft and reducing the efficiency of the runway. There is a kind of practice, construction of end-around taxiway at both ends of the runway, using natural terrain at the same time, considering the airport clearance requirements, shortening the distance of slide around. The landing aircraft enter the end-around taxiway through rapid exit taxiway and parallel taxiway, to slide into airport parking space, to avoid the aircrafts crossing the runway, to increase the service efficiency of the runway[6][7].

3.3. Apron capacity

Aprons of military and civil are facilities for parking planes, which function like parking lots for vehicles. In planning military airport airfield capacity, consider the plane from sliding in and out, quick deployment and so on, according to the plane dimension for the classification, in a park a variety of models is the sum of the number of aircraft targets, such as the apron size, turning radius, safe distance as constraint conditions, the use of mathematical software or simulation, get the best in aircraft parking way and could park the largest number of the plane.

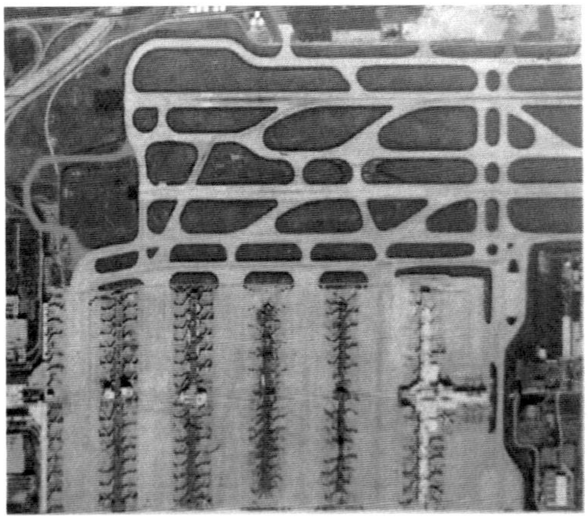

Figure 2. End-around taxiway of Atlanta airport

4. Displaced threshold When the conditions limited

The displaced threshold of the runway refers to the runway entrance not located at the end of the runway[8]. In some airports, after the expansion of the airfield area, the displaced threshold of the runway has been permanently adopted. The main reasons are as follows: with the increase of the mass of civil aircraft, the aircraft must have a relatively long runway to meet the requirements of take-off speed. When the aircraft is landing, the landing mass decreases, and the resistance can be increased by opening and increasing the wind resistance device and engine counter-injection. The length of landing runway does not need to reach the length of take-off runway. The displaced threshold of the runway reduces the length of the landing runway. The aircraft starts accelerating at the end of the runway during take-off until the take-off leaves the runway. Displaced thresholds of Hong-qiao airport are shown in Figure 3.The permanent displaced threshold of the runway has the following advantages.

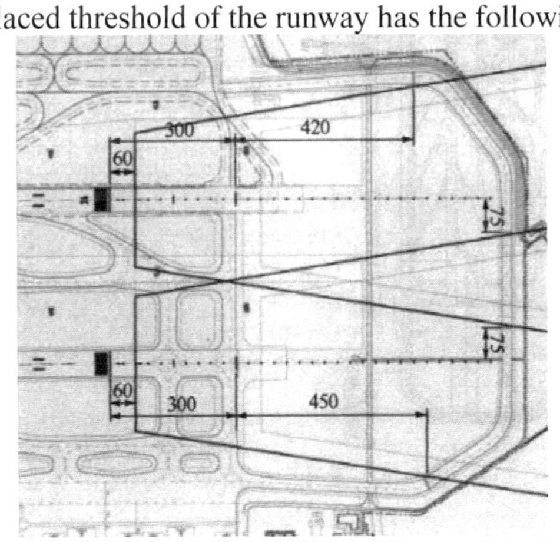

Figure 3. Displaced thresholds of Hong-qiao airport

4.1. Reduce land requisition

With the construction and development of the city, housing construction and municipal infrastructure around the airport arc also gradually improved, land resources and clearance resources are increasingly scarce, and land acquisition and demolition are increasingly difficult. It can be roughly estimated that if the runway of a single runway is extended 300 meters to one end, the land needs to be expropriated about 113,000 square meters. With the extension of the runway, the limit area of the airport obstacles expands and the number of obstacles to be dealt with increases. However, in the case of permanent inward movement of the runway entrance, when the runway is extended by 300 meters, only a small amount of land acquisition is needed, only additional obstacles in the restricted surface of the take-off runway obstacle are needed, and the approach and landing situation is not considered.

4.2. Reduce investment

The entry of the runway has been moved inward, and there are relatively few facilities to be renovated. It is mainly reflected in the following two aspects: navigation facilities, visual navigation AIDS. After the extension of the runway, when the runway entrance is set at the end of the new runway, omnidirectional beacon station, glide platform, rangefinder, beacon beacon station and other navigation stations should be rebuilt according to the construction requirements of the communication navigation station to meet the needs of aircraft navigation and the extension of the runway. With the extension of the runway, the original navigation AIDS cannot meet the requirements of aircraft approach and landing after the extension of the runway, so the approach AIDS lighting system must be reformed. If the runway entrance is moved inward, the navigation light system of the original runway can be modified to meet the use needs of the aircraft, and fewer facilities need to be built.

4.3. Avoid the expansion of the noise to some extent

Airport noise has been affecting the lives of the surrounding residents. With the increase of residents' awareness of environmental protection, people's requirements on the environment are getting higher and higher. Many airports are closed at a certain time during the night. After the runway is extended, if the plane lands at the end of the new runway, the flight path of the plane will change accordingly, and the influence area of noise will expand to some extent, which will affect more residents. If the runway entrance is moved inward, as the number of flights taking off and landing at the airport increases, the area affected by noise basically does not expand[9].

5. Conclusion

Based on the analysis of the airport operation management and the capacity of the airfield area, some suggestions on the planning and design of the airfield area are put forward. In view of the contradictions existing in the operation of military-civilian airports, solutions are put forward. Aiming at the problem of airfield capacity, the capacity of each part must be studied systematically. The key to solving the capacity problem is to make every part without "bottlenecks". Combined with the actual situation, the method of permanent displaced threshold of the runway is given, which is beneficial to solving the problem of land requisition, reducing investment and controlling noise, so as to meet the needs of airport reconstruction and expansion.

Acknowledgments

I want to thank Professor Shao Bin for his careful revision of this paper, who helped me point out many mistakes.

Reference

[1] CAI Liang-cai (2018). Airport planning and design. National Defense Industry Press.
[2] Code for master planning of civil airport (MH-2002). Civil Aviation Administration of China.
[3] LIU Wu-jun (2013). Airport planning. Shanghai Science and Technology Press.
[4] Technical specification for airport hourly capacity assessment (AP-93-TM-2017-01) [S]. Civil Aviation Administration of China.
[5] Aerodrome technical standards (MH5001-2013). Civil Aviation Administration of China.
[6] WANG Wei, LI Long (2016). Study on Operation Strategy of End-around Taxiway and Its Application. Journal of Highway and Transportation Research and Development, 33(6):119-122.
[7] XU Ting, HU Ming-hua, ZHAO Zheng(2015). Analysis of End-around Taxiway's Operational Benefits. Aeronautical Computing Technique, 45(4):112-115.
[8] CHEN Li (2017). Formulation of construction standard scheme of civil airport airfield reconstruction and expansion project without stopping navigation. Standardization in

China,07:136-137.
[9] ZENG Qing-hua, XU Zhi-sheng (2017). Study on Airport Site and Runway Layout Optimization to Reduce Aircraft Noise Impact. Environmental Impact Assessment,39(6):32-33.

IOP Conf. Series: Materials Science and Engineering **792** (2020) 012017 doi:10.1088/1757-899X/792/1/012017

Airport Rescue and firefighting Way in the 5g era

LI DaShuai[1*], Su Ruifang[1]

[1] Airport Planning and Design Institute, China Aviation International Construction and Investment Co., Ltd., Beijing, 100120, China

[*]Corresponding author's e-mail: 1073263692@qq.com

Abstract. With the maturity of the fifth generation of mobile communication technology (5G), the new technology associated with 5G will have a great impact on the way of Rescue and firefighting in airports. Based on the development of existing technologies in the 5G era, the personnel in the aircraft Positioning technology, real-time scenario monitoring, measurement and data transmission technology for drones, real-time simulation and prediction technology of the aircraft fire situation, and the combination of remote fire hydrants and unmanned RFF vehicles will greatly improve the efficiency and success rate of fire rescue, and reduce the risk to fire rescue workers.

1. Introduction

With the maturity of the fifth generation of mobile communication technology (5G), China is vigorously building the infrastructure of 5G network. 5G communication is based on its continuous wide area coverage, hot spots, high capacity, low power large connection and low latency, high reliable scene will bring a lot of technological changes. The innovation of these technologies has a profound impact on the social system. Specifically, airport fire rescue methods will undergo drastic changes. This article analyses the constraints of modern airport fire rescue methods and proposes that the author himself the concept of airport fire rescue in the 5G era.

2. Constraints in modern fire rescue methods

At present, the airport's rescue and fire is around the fire engine, after receiving an initial call, rescue and firefighting vehicles need to be equipped to start the vehicle and arrive at the scene. Upon arrival at the scene, rescue and firefighting personal needed to control the fire in the critical area for the first time, followed by the evacuation of trapped people inside the aircraft. The fire may lose control in a very short time after the plane is in danger. Therefore, the early arrival of the scene and evacuation of people is critical to the trapped people. This section mainly analyzes the constraints affecting rescue and fire protection.

2.1. Response time

One of the core indicators of airport fire rescue is response time, the operational objective of the rescue and firefighting service shall be to achieve a response time not exceeding three minutes to any point of each operational runway, in optimum visibility and surface conditions [1].

According to the definition of response time, there are two factors that have a greater impact on response time: one is the time required for personnel to wear equipment and the characteristics of the RFF vehicles itself; the other is visibility and ground conditions.

The modern airport fire rescue method mainly relies on RFF vehicles. At present, the Civil Aviation Administration of China stipulates that the airport fire station staff must not exceed one minute from receiving the police to leave the garage after receiving the signal. It takes almost one minute to start the RFF vehicles. At present, main foam RFF vehicles commonly used in domestic airports, for example Oshkosh New Striker 3,000 (6x6). This RFF vehicle can carry 11,356L of water, 1590L of AFFF concentrated foam, and the vehicle is rated the maximum total weight is 42,184kg, the maximum vehicle speed is 115kmp, and it can accelerate from static to 80kmp in 33s. RFF vehicles are extremely heavy due to the large amount of water loaded on it, which makes it difficult for the vehicle to accelerate and decelerate. The maximum driving speed of the RFF vehicles is limited. In the case of poor visibility such as rain, snow, haze and ground conditions, the driver can only reduce the speed of the vehicle, which also results in an extended response time.

2.2. Continuous water supplies

Calculating the total water required for aircraft firefighting usually uses the formula [2]:$\mathrm{Q} = Q_1 + Q_2$

Among them, Q—— the total water required

Q_1——the water for control of the fire in the practical critical area, and,

Q_2 ——the water required after control has been established and is needed for such factors as the maintenance of control and/or extinguishment of the remaining fire.

Under normal circumstances, the amount of water carried by the RFF vehicles can meet the requirements of Q_1, but it cannot meet the subsequent water quantity Q_2. RFF personnel also need to consider finding the nearest water source to replenish, and even return to the fire station to replenish. So, A continuous water supply is essential and is usually not available at all points [2].

2.3. Restrictive factors of rescue and firefighters themselves

The first batch of rescue and fire fighting vehicles that arrived at the scene of the accident was to cover the area affected by fuel in advance [2], and continue to spray fire extinguishing agents for more than 1 minute to control the fire in key areas. Evacuation or other tasks. If the response time of an airport is 3 minutes, it will take one minute for the first RFF vehicles to control the fire in a critical area, which means that in the most unfavourable situation, the people trapped inside the aircraft will not be able to get rescue within the first 4 minutes from external forces.

In addition, there is the safety risk of rescue and firefighters themselves. Firefighters formulate a fire suppression plan based on the fire situation of the aircraft, determine the location of the RFF vehicles, the path of the personnel close to the ignition point, and the target and direction of the fire extinguishing agent spray. From the perspective of protecting firefighters Considering that sometimes the best fire extinguishing position cannot be occupied, only a relatively reasonable position can be selected, thereby reducing the rescue efficiency.

3. New technologies and applications brought by 5g

Characteristics of 5G are continuous wide area coverage, hotspot high capacity, low power consumption, large connection, and low delay and high reliability. The 5G white paper released by IMT2020 (5G) clearly puts forward the indicators for 5G technology requirements [3]: User Experience The rate reaches 1Gbps, the peak rate reaches 10Gbps, the connection density is greater than 1,000,000/m^2, the end-to-end delay is 1ms, and the transmission reliability needs to be 99.9999%. In addition, 5G can provide communication services for users with a speed of 500km/h. 5G, the network can truly realize the interconnection of all things, the data can be transmitted at high speed and reliability, and the device can realize remote zero-delay control. The technologies that have a significant impact on airport fire rescue in the 5G era will include: the personnel in the aircraft Positioning technology, and real-time scenarios of drones Monitoring, measurement and data transmission technology, real-time simulation and prediction of aircraft fire conditions, and the combination of remote fire hydrants and unmanned RFF vehicles.

4. Airport fire rescue methods in the 5g era

4.1. The personnel in the aircraft Positioning technology

According to statistics, from 1998 to 2008, commercial jet aircraft worldwide suffered injuries due to accidents on the ground caused by controllable flight touchdowns, aircraft out of control (ground), and aircraft off the runway. The total number of people is 1,515. After the plane crash, due to the rapid spread of the fire, the fuselage was full of smoke, rescuers must wear breathing apparatus to enter the aircraft for search and rescue. The inhalation of high-temperature, smoke containing a large amount of toxic substances can cause in a short time People trapped inside the aircraft lost consciousness and the ability to move. So, improving search and rescue efficiency is a research topic for rescue researchers. The US FAA has explored the application of thermal imaging technology in the rescue field as early as 2017 [4]. But thermal imaging technology Due to the limitation of its principle, in the case of fire, the temperature situation is complicated, and once a person loses consciousness in an area close to his body temperature, the thermal imaging technology may not be found. At present, the application of thermal imaging technology in the field of fire protection is still in discovery and early warning of ignition points.

At present, there are relatively few studies on the rapid positioning of aircraft internal personnel. The personnel in the aircraft Positioning technology can play an active and effective role in Rescue and firefighting service. Based on the high reliability of 5G technology, the positioning system worn by passengers and crew members can be real-time the position of the sender can further reflect the physical condition of the person. The rescuer can fully understand the situation of the rescued person before entering the aircraft, which greatly improves the efficiency and accuracy of the rescue.

Table 1. Casualties of commercial jets worldwide from 1998 to 2008 [5].

Cause of the accident	Casualties
Plane out of control (air)	1926
Controllable aircraft touches the ground	961
Off track	408
System / component failure	539
Ground control service error	156
Plane out of control (ground)	146
Encounter wind shear	193

4.2. Drones and real-time simulation and prediction technology of the aircraft fire situation.

UAV's preliminary investigation, real-time scenario monitoring, measurement and data transmission during rescue. High-speed and reliable data transmission, remote zero-latency control of equipment, and ultra-high-speed scene service at 500km / h, which means that the aircraft is in danger in the air. At that time, drones can quickly take off and get a closer look at the situation of the aircraft. After the plane landed, multiple drones can be sent to the accident site, and multiple drones can transmit real-time scenes of the aircraft in the first time. Give the control and command center to make rescue commanders more efficient and accurate decision-making. When rescuers and equipment are in place, drones can play a greater role. Multiple drones cooperate on multiple angles of aircraft danger Shooting can overcome the problems of shooting dead angles and poor measurement accuracy of a single drone, so as to accurately measure the fire and fuel leakage of the aircraft. The monitoring center uses the measurement data to model the fire of the aircraft in real time, and can fully understand and even predict Fire Spread: The rescue plan will be more scientific and reasonable, which will greatly improve rescue efficiency and rescue success and reduce the dangers that rescuers encounter during the rescue process.

4.3. The combination of remote fire hydrants and unmanned RFF vehicles.

In the 5g era, remote fire hydrants have become a reality due to the remote zero-delay control of the equipment. A fire pipe network is set up at the airport, and remote fire hydrants are installed at the end of the fire network. It can be remotely controlled to rise to a certain height above the ground. Within the equipment protection distance, the angle and pressure of the water spray can be freely adjusted to extinguish the aircraft or add water to the RFF vehicles. Each remote fire hydrant can provide a certain range of protection. Reasonable configuration can make the entire airport be protected from all corners. Due to the existence of remote-controlled fire hydrants, RFF vehicles do not have to carry a large amount of water but can still be supplemented with water at any part of the airport. This allows RFF vehicles to greatly reduce their own weight. Compared to main foam cars, RFF vehicles of the 5G era It only needs to carry about 1200L of concentrated foam concentrate, foam mixing equipment and water cannon to extinguish the fire. This load is only equivalent to the weight of a pickup truck. The highest performance pickup truck can reach a maximum speed of 180km / h. It can accelerate from standstill to 100km / h in 15s. The performance in speed is far better than that of an RFF vehicle.

Unmanned driving will also be applied to RFF vehicles. The airport flight area is relatively simple compared to the urban road network. The driving environment is single, the interference factors are small, and the driving route is relatively fixed. IN this scenario, it is easy to train in rainy and snowy days, and on the road. The driving strategy on wet terrain, so as to get the fastest driving speed in various scenarios, which means that the theoretical optimal rescue efficiency can be achieved under any weather conditions. In fact, after removing the cockpit, the RFF vehicles also can get better acceleration performance and higher maximum speed. The low-weight body has lower requirements for the vehicle chassis, and even the existing ordinary off-road vehicle chassis can be competent, which can greatly reduce the technical threshold of RFF vehicles, and Directly reduce the cost. RFF vehicles technical threshold and lower cost can make the airport bear more unmanned RFF vehicles. Airports that were originally not equipped with RFF vehicles will also be equipped. This is a broad market and the scale effect of the market will further increase. Reduced the cost of RFF vehicles.

The unmanned RFF vehicles also mean that there is no need to carry firefighters, and it can be dispatched at the same time as receiving the instruction, saving 1 minute of time for response time. Before the unmanned RFF vehicles arrived at the scene, the drone had already sent the scene the situation is transmitted to the command center in real time, and the command center can use the time when the unmanned RFF vehicles rushed to the scene making a decision to determine the docking position and fire extinguishing method of the unmanned RFF vehicles. This brings two benefits: one is the unmanned RFF vehicles There are no rescue and firefighters on the scene, so it is not necessary to consider the safety situation of rescuers, so that they can implement more flexible and effective fire extinguishing tactics. The second is to control the fire in key areas before rescue and firefighters arrive at the scene. Time can be devoted to the evacuation and rescue of trapped people.

Compared with modern RFF vehicles, the combination of remote fire hydrants and unmanned RFF vehicles theoretically shows great advantages.

Table 2. Comparison of remote-control fire hydrant + unmanned RFF vehicles and RFF vehicles.

Contrast item	Unmanned RFF vehicles	Traditional RFF vehicles
RFF vehicle dispatch time	0	Nearly 1 minute
RFF vehicle acceleration performance	High	Low
Top speed of RFF vehicles	High	Low
RFF vehicles carrying amount of fire extinguishing agent	Many	Less
Fighting time of a single RFF vehicle	Long	Short
RFF vehicles cost	Low	High

5. Summary and conclusion.

The technology mentioned in this article is not a distant scientific fantasy. Indoor positioning technology has been commercialized, and some studies have shown that 5g combined with Beidou satellite navigation can already achieve centimetre-level positioning. The shape of the aircraft is single, and the structure It is simple and has less interference from electronic equipment, which is easier to

IOP Conf. Series: Materials Science and Engineering **792** (2020) 012017 doi:10.1088/1757-899X/792/1/012017

appear than the positioning technology in the building. The remote-control fire hydrant is essentially a combination of remote-control technology and telescopic boom, which has already been widely used in raising RFF vehicles. Machines and unmanned vehicles have been widely used in many fields. The combination of unmanned vehicles and RFF vehicles has not brought new technical challenges to the automotive industry, but instead has brought new opportunities: lower technology Thresholds, lower costs, and a broader market size. The combination of improved cost reduction and superior firefighting methods can give airports more incentive to upgrade existing fire equipment.

The new technology brought by the 5g era can change the definition of response time: the first person to arrive at the scene is no longer an RFF vehicle, but a drone; the RFF vehicles can reach it in 1 to 2 minutes under any weather and road conditions At the scene, before that, the remote-controlled fire hydrant has been ready to add water to the RFF vehicles or directly spray water to extinguish the fire; the firefighters know the situation of the aircraft at the rescue site and can develop a rescue plan for each trapped person. It is not explored in smoke. This is the 5g era airport fire rescue method envisaged by the author. Advances in technology will not only change the rescue method, but will also profoundly affect the construction of the airport fire rescue system. In the future, the author will also make corresponding research on the transformation of the rescue system.

References

[1] International Civil Aviation Organization, July 2018.Civil Aviation-Annex 14 Aerodromes Volume I Aerodrome Design and Operations;
[2] International Civil Aviation Organization, 2015, Airport Services Manual Part 1 — Rescue and Firefighting Fourth Edition
[3] IMT-2020 (5G) Promotion Group, 2017. 5G economic and social impact. White Paper.
[4] Matthew, S., Jonathan T., *Jack K. (2017) Thermal Imaging for Aircraft Rescue and Fire Fighting Applications, Federal Aviation Administration, Atlantic City.
[5] Cheng F. (2017) Based on the Human Factors of Evaluative Research about the Ability of Student Pilot Dispose Flight Special Situation. Civil Aviation Flight University of China, Guang Han.

Reliability analysis of highway network in China based on complex network theory

Hui-di Jia[1*], Qian Li [2], Guang-Jing Zhou[1] and Ran Zuo[1]

[1] School of Traffic and Transportation, Beijing Jiaotong University, Beijing, Province, 100000, China

[2] Integrated Transportation Research Center, China Academy of Transportation Sciences, Beijing, 100000, China

[*]Corresponding author's e-mail: huidijia@fox.mail

Abstract. In this paper, complex network theory is introduced to analyze the Chinese highway network. 72 cities were extracted from highway information in China and topological network was established based on L space. Then degree distribution, betweenness, clustering coefficient and other network characteristics of Chinese highway network were analyzed. Through computer simulation, the reliability of the Chinese highway network was carried out. The results show that the highway network in China is relatively complete, the connection between cities is convenient.

1. Introduction

In the past 10 years, network research has gradually attracted people's interest. And complex network theory is applied to many subjects, such as physics [1], biology [2] and economics [3]. Transportation network supports the movement of people, goods and information, so for us to know its structure and performance is critical. Wang uses a complex network approach to examine the network structure and nodal centrality of individual cities in the air transport network of China (ATNC) [4]. Dai and Liu capture the main topological and spatial changes of Asian air transport network (SAAN) [5]. Mohmand and Wang contributes a weighted complex network analysis of travel routes on the national highway network of Pakistan [6].

Zhang and Peng modelled and simulated the Taian city bus lines and concluded that the Taian city network had small-world character [7]. Wang and Li study the public transport network of Shenyang as a complex network. On the basis of a large number of statistical data, the characteristics of complex network in Nanjing are researched through Space L and Space P methods [8]. Xu and Zu studies the public transport networks of 330 cities in China based on the space P concept [9]. Li, Zhu and Zhou study the urban public transport network (UPTN) in Nanjing, and the empirical data indicates that it is a small-world network [10]. Gu, Li and Qin analyzes the spatial structure of Chinese railway network with indicators of degree distribution, the clustering coefficient, the average path length and centralization [11].

In recent years, China's expressways have been developing faster and faster. By the end of 2018, the total length of expressways in China had exceeded 140,000 kilometers, ranking first in the world. However, few studies have analyzed high-speed networks at the national level. To this end, this paper attempts to analyze the characteristics of the Chinese highway network based on complex network theory.

The innovation of this paper lies in the establishment of the highway network topology of China's transportation hub cities and the analysis and evaluation of the highway network based on the complex network theory.

The rest of this paper is organized as follows. Section 2 discusses the data of this study and Section 3 discusses the network construction methodology. In Section 4, we will analyze the topological properties of Chinese highway network systematically. Finally, the conclusions are given in Section 5.

2. Methodology

The methodology of the study basically includes various steps, such as (1) data collection of Chinese highway network; (2) network construction of Chinese highway network; (3) analysis of Chinese highway network. And (3) is comprised of three major components: degree and betweenness; clustering coefficient and distance distribution; network reliability.

2.1. Degree and betweenness

Node degree refers to the number of edges associated with the node. It is also known as correlation degree. The larger the degree of a node, the more important it is. So the importance of the city can be measured by the degree between two nodes. The mathematical expression is as equation (1):

$$k_{v_i} = \sum_{j}^{N} x_{ij} \tag{1}$$

Among them, N is the number of total nodes, k_{v_i} represent the degree of node v_i. If v_i is connected with v_j, then $x_{ij} = 1$, or $x_{ij} = 0$.

By counting how many nodes have each degree, we form the degree distribution $p_{deg}(k)$, defined by the fraction of nodes in the graph with degree k.

The node betweenness is defined as the proportion of the total number of shortest paths in the network to the total number of shortest paths. The mathematical expression is as equation (2):

$$B_{v_i} = \sum_{v_j \neq v_l \neq v_i} \left[\frac{N_{v_j v_l}(v_i)}{N_{v_j v_l}} \right] \tag{2}$$

Where, v_j and v_l are two nodes in the network that are not adjacent. $N_{v_j v_l}(v_i)$ is the number of the shortest path between node v_j and v_l passing through the node v_i. $N_{v_j v_l}$ is the total number of minimum paths between nodes v_j and v_l.

The nodes with a large number of media are often the key nodes of the optimal transport route of cargo or the optimal route of passenger journey.

2.2. Clustering coefficient and distance distribution

The clustering coefficient and distance distribution are important parameters to describe the characteristics of a network.

The coefficient of agglomeration is used to describe the degree of clustering between the vertices in a graph. Specifically, the degree to which the adjacent points of a point are connected to each other. The clustering coefficient C of a network is defined as the average of the clustering coefficient C_i of all nodes in the network. C_i is defined as follow:

$$C_{v_i} = \frac{2e_{v_i}}{s_{v_i}\left(s_{v_i} - 1\right)} \tag{3}$$

Among them, e_i is the number of edges that actually exist between the s_{v_i} neighbor nodes of node v_i. The denominator is the largest possible number of edges. The higher the average aggregation coefficient, the closer the connection between nodes is.

The distance d_{ij} between node v_i and node v_j is defined as the number of edges contained in the shortest path connecting these two nodes. The average path length L of the network is defined as the average of the distance between any two nodes. L is defined as follow:

$$L = \frac{2}{N(N-1)} \sum_{1 \le i < j \le N}^{N} d_{ij} \tag{4}$$

Among them, N is the number of total nodes. The distance between any two nodes reflects the minimum number of points or edges needed to get from one point to the other. The smaller the distance, the better the connectivity.

2.3. Network reliability

There are many ways to measure network reliability and damage resistance, Xing, Lu and Chen took the Shanghai Metro Network (SMN) as an example to investigate vulnerability of a weighted metro network in responding to random failures as well as malicious attacks [12]. Hui, Cai and Rui studied the random failure and selective attack of Lanzhou public transport complex networks [13]. Ruan and He proposes a CA-SVM based Monte Carlo approach to evaluate network reliability [14].

In this paper, the global efficiency of the network is used to measure the reliability of CHN. The global efficiency of the network is defined as follows:

$$E = \frac{2}{N(N-1)} \sum_{1 \le i < j \le N}^{N} \frac{1}{d_{ij}} \tag{5}$$

3. Network construction of the highway network

In this study, according to the 13th five-year plan for the development of the modern comprehensive transportation system, we choose representative 72 cities to evaluate the highway network. The data is collected from Baidu map.

Generally, traffic network can be defined as P space, R space, L space and C space. This study is based on L space. In the CHN, a city is considered as a node, and an edge represents a connected route between two cities. Highways are bidirectional, so we consider CHN as an undirected network. The typical topology of CHN is shown in Figure 1.

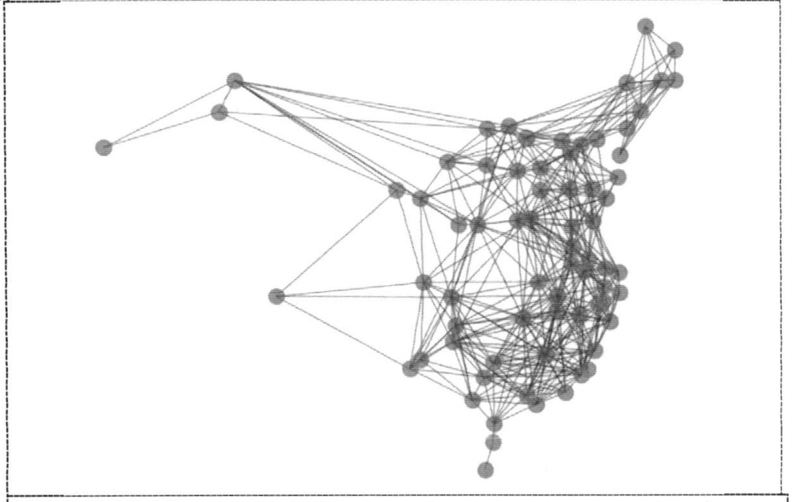

Figure 1. Topological structure of Chinese highway network.

4. Characteristics of highway network in China

4.1. Degree and betweenness

IOP Conf. Series: Materials Science and Engineering **792** (2020) 012018 doi:10.1088/1757-899X/792/1/012018

Figure 2 shows the degree distribution of the Chinese highway network. According to the entire highway network, the average node degree is 12.2. There are 49% of nodes with degree value bigger than 12.2. Zhengzhou and Jiujiang have the highest degrees, which 23. The degree value of most nodes is concentrated between 10 and 15, and a few nodes have lower or higher degree value, that is, Chinese highway network is relatively complete.

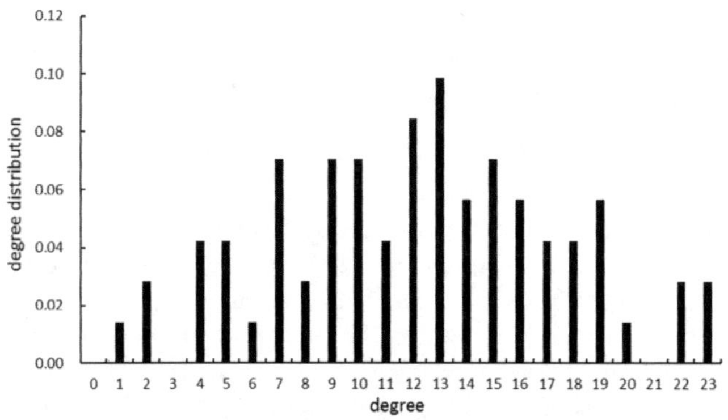

Figure 2. Degree distribution.

In order to measure the importance of each station node in the Chinese highway network more comprehensively, the node betweenness is calculated in this paper.

Table 1 shows the top 15 stations in degree and medium betweenness. We find there is a certain correlation between the betweenness and degree, and the correlation coefficient of the two is 0.59. In other words, the betweenness of nodes with low degree value is also relatively lower, while the betweenness of nodes with high degree value is often higher.

Table 1. Top 15 stations in degree and medium betweenness.

City	Degree	City	Degree
Zhengzhou	23	Wuhan	0.3145
Jiujiang	23	Zhengzhou	0.299
Changsha	22	Shijiangzhuang	0.2823
Ganzhou	22	Nanjing	0.2639
Hefei	20	Changsha	0.2636
Chongqing	19	Nanchang	0.2274
Xian	19	Jinan	0.2153
Jinan	19	Guangzhou	0.184
Guiyang	19	Shenyang	0.1647
Nanjing	18	Beijing	0.161
Nanchang	18	Chongqing	0.1311
Shangrao	18	Lanzhou	0.116
Tianjin	17	Tianjin	0.1094
Luoyang	17	Fuzhou	0.1076
Guilin	17	Xian	0.0773

4.2. Clustering coefficient and distance distribution

The cumulative clustering coefficient frequency diagram of the Chinese highway network is shown in figure 3. More than 67% of the node clustering coefficient over 0.5, the average clustering coefficient is 0.55, showing high aggregation, this shows that the cities in Chinese highway network have establish close contact, and with the increasingly highway construction, the network average clustering coefficient will increase. This means that the relationship between node will be more and more closely.

IOP Conf. Series: Materials Science and Engineering **792** (2020) 012018 doi:10.1088/1757-899X/792/1/012018

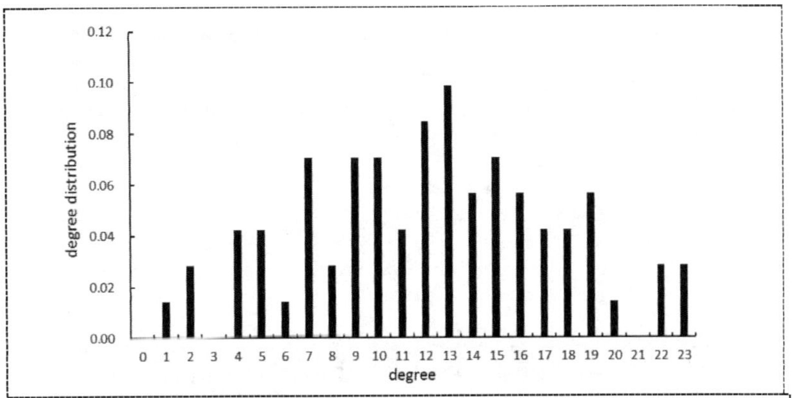

Figure 3. Degree distribution.

By studying the node distance of the Chinese highway network, we found that the average path length is 2.48. This means that it takes an average of 2-3 cities to get from one city to another, which is a smaller average path length. Since the network has a high clustering coefficient and a short average path length, the Chinese highway network constructed in this paper is a small-world network.

Table 2 lists the top 20 nodes with the longest average shortest path length of all nodes. Due to their geographical location on the edge of China's territory, these nodes are less connected to other cities and are far away from other cities.

Table 2. The first 20 nodes with the longest average shortest path length.

Station	Length	Station	Length
Sanya	4.507042	Shenyang	2.760563
Haikou	3.535211	Xining	2.746479
Dalian	3.492958	Yantai	2.676056
Qiqihaer	3.450704	Quanzhou	2.647887
Jilin	3.422535	Ningbo	2.633803
Keshen	3.408451	Baotou	2.619718
Kuerle	3.140845	Shantou	2.619718
Yingkou	2.971831	Qinhuangdao	2.591549
Haerbin	2.887324	Zhenjiang	2.591549
Lasa	2.774648	Qujing	2.591549
Sanya	4.507042	Shenyang	2.760563
Haikou	3.535211	Xining	2.746479
Dalian	3.492958	Yantai	2.676056
Qiqihaer	3.450704	Quanzhou	2.647887
Jilin	3.422535	Ningbo	2.633803

4.3. Network reliability

Through computer simulation, random attacks and deliberate attacks were conducted on the nodes of the Chinese highway network, in which the deliberate attacks of the nodes were carried out according to the maximum degree of node attack and the maximum number of intermediate node attack.

According to the figure 4 and figure 5, we can find that when attacking nodes, deliberate attacks are more destructive than random attacks, but the difference is not obvious. One reason is that the Chinese highway network is multi-core, and even a deliberate attack would not immediately bring down the global network. On the other hand, the network has a large connection coverage, so the connection path between cities will not be completely destroyed because of one or two attacks.

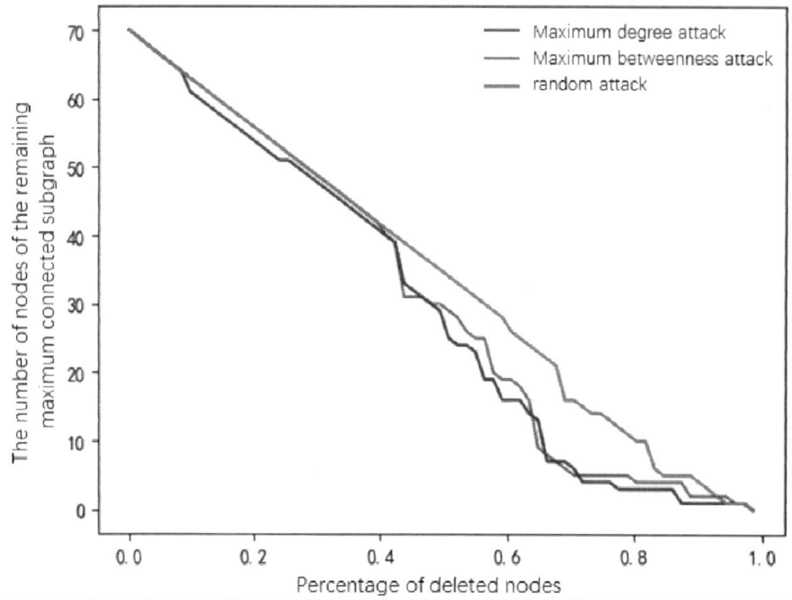

Figure 4. The change curve of the maximal connected subgraph of highway network under attack.

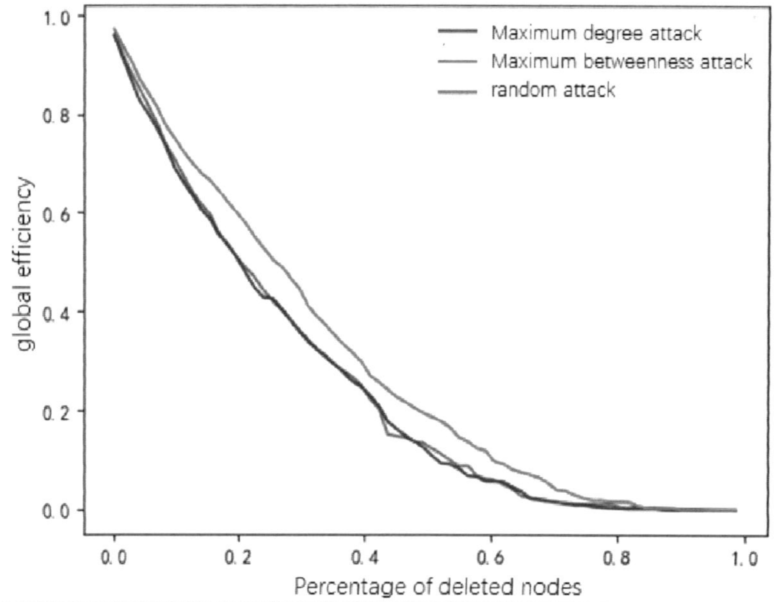

Figure 5. The curve of highway network efficiency under deliberate attack.

5. Conclusion

In this paper, based on the research method of complex network, the network characteristics of the established Chinese highway network is analyzed. It is found that the Chinese highway network is a small-world network. The average clustering coefficient is 0.55 and the average path length is 2.48. The betweenness is positively correlated to the degree. In the face of attack, deliberate attack is more destructive than random attack, but the difference is not obvious. It indicates that Chinese highway network connection coverage is relatively large.

IOP Conf. Series: Materials Science and Engineering **792** (2020) 012018 doi:10.1088/1757-899X/792/1/012018

A sound transportation network can promote economic development and strengthen exchanges between regions. In this paper, the analysis of the structure characteristics and reliability of the highway network is helpful to the optimization of highway transportation. This also provides a basis for the further improvement and maintenance of the expressway network.

Acknowledgments
The research is supported by China Academy of Transportation Sciences (Project ID: 20196661184). The author wishes to thank the agency.

References
[1] E. Cervi, A.; Cammi, E. Zio. (2019) A new approach for nuclear reactor analysis based on complex network theory. Progress in Nuclear Energy, 112:96-106.
[2] K McGarry.; S McDonald. (2018) Complex network theory for the identification and assessment of candidate protein targets. Computers in Biology and Medicine, 97:113-123.
[3] G Cao, Q Zhang, Q Li. (2017) Causal relationship between the global foreign exchange market based on complex networks and entropy theory. Chaos, Solitons & Fractals, 99:36-44.
[4] Wang J, Mo H, Wang F, et al. (2011) Exploring the network structure and nodal centrality of China's air transport network: A complex network approach. Journal of Transport Geography, 19(4):0-721.
[5] Dai L, Derudder B, Liu X. (2018) The evolving structure of the Southeast Asian air transport network through the lens of complex networks, 1979–2012. Journal of Transport Geography, 68:67-77.
[6] Mohmand Y T, Wang A. (2013) Weighted Complex Network Analysis of Pakistan Highways. Discrete Dynamics in Nature and Society, 2013:1-5.
[7] Zhang L, Peng L, Zhang Y, et al. Urban Bus Transport Network Optimization from Complex Network. In: 2012 Proceedings of International Conference on Modelling, Identification and Control, Wuhan, Hubei, China. pp. 1159-1164.
[8] Wang D, Li X. (2013) Analysis of the Public Transport Network of Shenyang as a Complex Network. Applied Mechanics and Materials, 433-435:1370-1373.
[9] Xu Q, Zu Z, Xu Z, et al. (2013) Space P-Based Empirical Research on Public Transport Complex Networks in 330 Cities of China. Journal of Transportation Systems Engineering and Information Technology, 13(1):193-198.
[10] Li, P., Zhu, Z T., Zhou, J., Ding, J Y., Wang, H W., & Wei, S S.. (2009). Performance analysis of public transport systems in nanjing based on network topology. Complex Sciences, First International Conference, Complex 2009, Shanghai, China. pp. 254-264.
[11] Gu X, Li D, Qin L. (2012) Spatial Structural Characteristics of Chinese Railway Passenger Network Based on Complex Network Theory. In: 2012 National Conference on Information Technology and Computer Science. Lanzhou, China. 750-753.
[12] Xing Y, Lu J, Chen S, et al. (2017) Vulnerability analysis of urban rail transit based on complex network theory: a case study of Shanghai Metro. Public Transport, 9: 501–525.
[13] Hui-Fang F, Cai-Hong L I, Rui W. (2016) Vulnerability Study for Public Transport Network of Valley City:Case of Lanzhou. Journal of Transportation Systems Engineering and Information Technology, 16(1):217-222.
[14] Ruan Y P, He Z. (2013) A CA-SVM Based Monte Carlo Approach for Evaluating Complex Network Reliability[J]. Springer-Verlag Berlin Heidelberg, 19: 609-617

AUTHOR INDEX